To Vics!

With my thanks & Congratulations!

02/11/22

The Space Value of Money

Armen V. Papazian

The Space Value of Money

Rethinking Finance Beyond Risk & Time

Armen V. Papazian
King's College
University of Cambridge
Cambridge, UK

ISBN 978-1-137-59487-7 ISBN 978-1-137-59489-1 (eBook)
https://doi.org/10.1057/978-1-137-59489-1

This Palgrave Macmillan imprint is published by the registered company Springer Nature Limited
The registered company address is: The Campus, 4 Crinan Street, London, N1 9XW, United Kingdom

E pur si muove…[1]

[1] 'And yet it moves', refers to the Earth and is attributed to Galileo Galilei (1564–1642).

To my children,
Dare to imagine fearlessly, love unconditionally, and create responsibly.

Preface

Finance, just like all human inventions, is designed and structured in human reason and imagination first. It is actualised, institutionalised, and digitised, in some shape or form, perfectly or imperfectly, only after its purpose, principles, tools, and models have been defined, debated, and negotiated in society.

Digitisation does not automatically imply improvement. An unfair and unequal process can remain so after digitisation. The architecture of our markets, the procedural mechanisms behind money creation, the principles and equations of a valuation model change not when they are digitised, but when they are reinterpreted in their fundamental assumptions, and are rebuilt upon an entirely new value framework. In the age of digital transformation, we are more than ever exposed to the risk of digitising confusion, and reinforcing suboptimal frameworks, structures, and models.

As the world becomes more aware of the devastating impact of climate change and recognises that we have littered every environment we have come to touch—debris in orbit, carbon in air, plastic in oceans, waste in rivers and on land—the necessity for change has become a mainstream agenda. Whatever the perceived and real levels of commitment, the growth in sustainable finance is a testimony to this parallel and simultaneous transformation.

This book is a theoretical treatise on sustainability in finance and aims to contribute to the debate. The main argument and purpose are straightforward. If we are to ensure an effective transition, and a long-term change in our trajectory, we must integrate sustainability into the core principles and equations of finance, in theory and practice. I offer an approach and a set of equations that can achieve such an objective.

To entrench sustainability into finance, we must introduce space, as an analytical dimension and our physical context, into our models and equations. The main insights and ideas presented aim to provide a roadmap to an entirely new type of finance, where space and our impact, in it and on it, are an integral element of our value models. The book proposes a change in the logic of the value of money as taught and applied in finance theory and practice, in academia and industry. It entrenches our spatial responsibility into our value paradigm heretofore entirely focused on risk and time.

The ideas and analysis proposed in this book have been in development for more than a decade. I have, throughout the last many years, starting in 2009, at different occasions and through different mediums and publications, shared different elements of the key concepts and metrics. Many of the pieces of the puzzle that I have previously discussed have been changed and transformed through the completed framework and analysis. I expect that these propositions will continue to evolve through the collective debate that improves and refines all ideas.

Whether a layperson or finance student or scholar, or practitioner, when considering the concepts presented and discussed in this book, I urge you to uphold an optimistic interpretation of human nature, and to recognise the immense creative power we possess as individuals, and as a global collective. Indeed, the incredible innovations in technology over the last many decades should be enough evidence that the human mind is capable of great feats, and it is time to expect innovative improvements in finance theory and practice as well.

The principle and metrics I propose and discuss in this book are not statistically tested, yet, because the aim is not to look for correlations in the past—a past from which our future must be so very different. The statistical testing will become possible only after the principle and metrics are adopted in theory and practice. While the equations proposed can be applied immediately, my focus is on making the case for a transformed value framework that takes us beyond risk and time, beyond the risk-averse return maximising individual investor.

I introduce humanity and the planet as the two stakeholders yet to be formally accounted for in our financial value equations. This is important given that the analytical framework of the discipline has been built around the risk-averse investor concerned with the risk and time value of expected cash flows. A framework that has brought us to the edge of an existential crisis with potentially catastrophic implications for our ecosystem.

We face the pressing need to transform money and finance, a discipline and industry that have, sometimes unwittingly and sometimes intentionally, damaged the very fabric of our ecosystem and thwarted the evolution of our species. Indeed, the root cause of our current predicament is not the carbon in our air, the plastic in our oceans, the radioactive waste on land, the sewage and garbage in our rivers, or the debris in orbit, but the lack of human responsibility, a monetary architecture that absolves it, and a discipline that has legitimised both.

If we are to truly change course and secure the health of our home planet, and the future of our children, we must reimagine the value of money and the institutional structures that create it. We face an evolutionary choice that will determine our survival and the sustainable expansion of human productivity on this planet and beyond.

United Kingdom
22 February 2022

Armen V. Papazian

Acknowledgments

I owe thanks and gratitude to a number of institutions and individuals. I have received moral, financial, and intellectual support during the research and writing phases of this book, and I would not have been able to complete this work without it. I am grateful and thankful to:

King's College Cambridge University, for providing me with an intellectual home, and continuously supporting and inspiring my work.

Judge Business School Cambridge University for giving me the support and the platform to express the early insights that are discussed in this book.

Tech Nation, for providing me with the support and sponsorship that allowed me to continue working on this book.

Hermes Investment Management, recently Federated Hermes International, for providing research funding during the early stages of the writing of this book.

Finoptek, including founding partners and shareholders, for their faith and support throughout the last many years.

Value Xd, including shareholders, colleagues, suppliers, and friends, for their continuous support.

I owe special thanks to my parents, my mother Aline and late father Varant Papazian, your love, courage, and humanity have inspired me throughout my life; to my partner and friend, Taline Artinian, you have made everything interesting and possible; to my wonderful children, Aren, Nayira, Roupen, and Vicken, you have given meaning to it all; to my sisters, Taline, Garine, Any, and their families, your continuous support has been invaluable; to my late grandparents, Dr. Papken and Arminée Papazian, and Aram and Sato Amassian, their bold kindness and example have been a guiding light.

I owe special thanks to Tula Weis for her patience and strategic support, she has played a central role in bringing this book to light, to Faith Su for reviewing the manuscript and making sure everything is as it should be, and to Pete Baker for recognising the theoretical and practical relevance of my work and making this publication possible.

I owe special thanks to Dr. Keith Carne, Prof. Gishan Dissanaike, Lt Col. Peter Garretson, Daud Vicary, Dr. Pascal Blanqué, Domenico del Re, Dr. Salvatore Russo, Prof. Christine Hauskeller, Dr. Saker Nusseibeh, Giotto Castelli, Adrian Webb, and Dr. Jonathan Bonello, for their support, for reviewing the early drafts of this book, for feedback and discussions, for providing reviews.

I am grateful to the following individuals for their direct and/or indirect contributions, recently or in the past:

Prof. Dame Sandra Dawson, Dr. Mark Carney, Prof. Geoff Meeks, Prof. Arnoud De Meyer, Prof. Ha-Joon Chang, Dr. Robin Chatterjee, Dr. Jose Gabriel Palma, Prof. Tony Lawson, Prof. Geoffrey Hodgson, Prof. Pierre-Charles Pradier, Prof. Peter Nolan, Prof Shailaja Fennell, Prof. Richard Barker, Dr. Rachel Armstrong, Dr. Richard Obousy, Kelvin F. Long, Dr. Ian J. O'Neill, Amalie Sinclair, John Lee, Prof. Pier Marzocca, Prof. Dirk Schulze-Makuch, Prof. Joseph Miller, Charles Radley, Giorgio Gaviraghi, Eric Klein, David Brin, Prof. Edward Guinan, Dr. Cathy W. Swan, Prof. Peter A. Swan, Dr. Jose Cordeiro, Dr. Robert L. Frantz, Prof. Weilian Su, Dr. Mae Jemison, Dr. David Livingston, Michael Laine, Dr. Eric Davis, Marc G. Millis, Paul Gilster, John Davies, Dr. Andreas Hein, Robert Swinney, Patrick Mahon, Robert Kennedy III, Angelo Genovese, Prof. Gregory Matloff, Mike Mongo, Prof. Mauricio Talebi Gomes, Dr. Riccardo Vitale, Dr. Tara Cornish, Dr. Tarek Mady, Dr. Shiyun Wang, George Littlejohn, Tony Manwaring, Richard Spencer, Prof. Harold Chorney, Dr. Seishi Kimura, and Dr. Everett Price.

I am also grateful to the many colleagues, students, friends, and family who have contributed to the wealth and depth of my learning and experiences over the years. I have learned so much from so many.

While all are to be thanked, mistakes remain my own.

Contents

About the Author

Armen V. Papazian is a financial economist, a visionary thinker and innovator. His research work on sustainability and money have been on the cutting-edge of theoretical finance, and his industry work has led to the creation of new markets, products, and a software platform that reimagines analytics.

He is the first winner of the Alpha Centauri Prize for his work on money mechanics for space (2013). He was a finalist for the Finance for the Future Awards for his work on Space Value of Money (2016). His innovative analytics platform was selected as one of top 33 UK Tech Startups (2019), listed amongst top innovative technologies, and shortlisted for various awards (2022).

He is a former stock exchange executive, an investment banker, a lecturer in finance, a consultant, and a researcher. He is the author of numerous articles and thought leadership contributions across professional and academic publications and global media. He earned his Ph.D. at the Cambridge University Judge Business School, King's College Cambridge.

An eclectic thinker with a passion for real life application and industry, Armen is a radical humanist whose work and ideas have served and contributed to a more holistic interpretation of our place in the universe, and the role of finance in securing the resources needed for the sustainable expansion of our creative reach.

Our financial imagination is as important as our technological imagination when it comes to extending our reach into the cosmos.

Armen V. Papazian, Starship Congress, 2013

Abbreviations

AI	Artificial Intelligence
APF	Asset Purchase Facility
APT	Arbitrage Pricing Theory
ASTP	Advanced Space Transportation Program
AT	Algorithmic Trading
BIM	Biodiversity Impact Metric
CAA	Climate Ambition Alliance
CAPM	Capital Asset Pricing Model
CBD	Convention on Biological Diversity
CCC	Climate Change Committee
CDE	Carbon Dioxide Equivalency
CDO	Collateralised Debt Obligations
CDP	Carbon Disclosure Project
CDSB	Climate Disclosure Standards Board
CE	Credit Easing
CGFI	UK Centre for Greening Finance and Investment
CGFI-SFI	UK Centre for Greening Finance and Investment, Spatial Finance Initiative
CISL	Cambridge Institute for Sustainability Leadership
COP26	UN's 26th Conference of the Parties
CPI	Climate Project Initiative
DCF	Discounted Cash Flow
DDM	Dividends Discount Model
DOJ	Department of Justice
EA	Environmental Agency

EELV	Evolved Expendable Launch Vehicles
EIO-LCA	Economic Input Output—Life Cycle Assessment
ELE	Extinction Level Event
EPA	Environmental Protection Agency
ESA	European Space Agency
ESG	Environmental, Social, Governance
ETC	Energy Transition Commission
FCA	Financial Conduct Authority
FCF	Free Cash Flows
FCFE	Free Cash Flows for Equity
FCFF	Free Cash Flows for Firm
FSB-TCFD	Financial Stability Board—Task Force on Climate-related Financial Disclosures
GFANZ	Glasgow Financial Alliance for Net Zero
GDI	Green Design Institute
GEO	Geostationary Orbit
GGM	Gordon Growth Model
GHG	Greenhouse Gas
GRI	Global Reporting Initiative
GSV	Gross Space Value
GTP	Global Temperature Potential
GWP	Global Warming Potential
HFT	High Frequency Trading
HRC	Habitat Replacement Costs
IBRD	International Bank for Reconstruction and Development
IEA	International Energy Agency
IFRS	International Financial Reporting Standards
IIED	International Institute for Environment and Development
IPBES	Intergovernmental Science-Policy Platform on Biodiversity and Ecosystem Services
IPCC	Intergovernmental Panel on Climate Change
IRR	Internal Rate of Return
ISS	International Space Station
ISSB	International Sustainability Standards Board
LCA	Life Cycle Assessment
LEO	Low Earth Orbit
LTV	Loan to Value
MBS	Mortgage-Backed Securities
MEO	Medium Earth Orbit
NASA	National Aeronautics and Space Administration
NLP	Natural Language Processing
NOPLAT	Net Operating Profit Less Adjusted Taxes
NPV	Net Present Value
NSV	Net Space Value

NZAM	Net Zero Asset Managers Initiative
NZE	Net Zero Emissions
OECD	Organisation for Economic Co-operation and Development
PA	Paris Agreement
PCN	Public Capitalisation Notes
PRA	Prudential Regulation Authority
PRI	Principles of Responsible Investment
QE	Quantitative Easing
ROIC	Return on Invested Capital
RTZ	Race to Zero
SASB	Sustainability Accounting Standards Board
SBTi	Science Based Targets initiative
SDGs	Sustainable Development Goals
SF	Sustainable Finance
SIIT	Social Impact Investment Taskforce
SVM	Space Value of Money
TCFD	Task Force on Climate-related Financial Disclosures
TCFD-PAT	TCFD, Portfolio Alignment Team
TCRE	Transient Climate Response to Cumulative CO_2 Emissions
TNFD	Task Force on Nature-related Financial Disclosures
UNEP	United Nations Environment Program
UNEPFI	United Nations Environment Program Finance Initiative
UNFCCC	United Nations Framework Convention on Climate Change
UNGC	United Nations Global Compact
UNPRI	United Nations Principles of Responsible Investment
VE	Value Easing
VRF	Value Reporting Foundation
WACC	Weighted Average Cost of Capital
WBG	World Bank Group
WWF	World Wildlife Fund

List of Figures

List of Tables

List of Charts

1

Introduction

> The difficulty lies, not in the new ideas, but in escaping from the old ones, which ramify, for those brought up as most of us have been, into every corner of our minds.
>
> **John Maynard Keynes**, The General Theory of Employment, Interest and Money, 1936

> The most important men in town would come to fawn on me,
> They would ask me to advise them, like a Solomon the Wise.
> 'If you please, Reb Tevye…'
> 'Pardon me, Reb Tevye…'
> Posing problems that would cross a rabbi's eyes…
> And it won't make one bit of difference if I answer right or wrong,
> When you're rich, they think you really know.
>
> **Tevye**, If I Were a Rich Man, Fiddler on the Roof, 1964[1]

'It could be structured by cows, and we would rate it'. This is an instant message sent by a Standard & Poor's employee in April 2007, reported by the Wall Street Journal and revealed through the investigations that followed the 2008 collapse of Bear Stearns and Lehman Brothers. This quote was widely cited at the time and is referenced in the US House Hearing that occurred

[1] The song was written by Sheldon Harnick and Jerry Bock inspired by the 1902 monologue by Sholem Aleichem 'If I Were Rothschild' (Aleichem 1949).

A. V. Papazian, *The Space Value of Money*,
https://doi.org/10.1057/978-1-137-59489-1_1

on the 22nd of October 2008, on Credit Rating Agencies and the Financial Crisis (Lucchetti & Burns, 2008; US Congress, 2008). Indeed, these investment banks had investment grade ratings until five days before their bankruptcy. Some rating agencies downgraded them only one business day before their collapse.

This led to numerous lawsuits against the rating agencies. In February 2013, the US Department of Justice filed a $5 billion lawsuit against S&P for fraud (US DOJ, 2013). In the same spirit, numerous multibillion lawsuits were filed against notable investment banks. Amongst others, Goldman Sachs agreed to pay more than $5 Billion in relation to its sale of residential Mortgage-Backed Securities (US DOJ, 2016).

What would happen if our ESG (Environmental, Social, Governance) ratings turned out to be as irrelevant, or as ineffective, as these credit ratings proved to be? Indeed, many of the rating agencies and financial institutions involved in the 2008 crisis are now active providers and consumers of ESG ratings—an integral part of their sustainability strategy and commitment. While the collapse of Mortgage-Backed Securities and Collateralised Debt Obligations brought the entire global financial system to a halt and led to trillions of dollars of losses across markets and economies (Lewis, 2010; Lowenstein, 2010), getting our ESG ratings wrong or missing our sustainability targets can have serious and far reaching consequences on our ability to address the climate crisis and the ongoing degradation of our environment. Moreover, no amount of debt-based bailouts would be able to fix the breakdown of our ecosystem.

The impact of climate change and the scientific evidence confirming human responsibility have been overwhelming (IPCC, 2013, 2018, 2021, 2022). In its most recent report, IPCC (2022) summarises the challenge:

> Human-induced climate change, including more frequent and intense extreme events, has caused widespread adverse impacts and related losses and damages to nature and people, beyond natural climate variability. Some development and adaptation efforts have reduced vulnerability. Across sectors and regions, the most vulnerable people and systems are observed to be disproportionately affected. The rise in weather and climate extremes has led to some irreversible impacts as natural and human systems are pushed beyond their ability to adapt (IPCC, 2022, 7)

Since the 2015 Paris Agreement (UNFCCC, 2015), we have witnessed a fundamental change in public discourse. The targets of the Paris Agreement, to keep world temperature increases below 2 °C above pre-industrial levels and ideally limit the temperature increase to 1.5 °C, have become a global priority. A multitude of initiatives in finance and industry are raising awareness and readjusting mainstream focus on the subject. International agencies have contributed detailed and complex analysis on what it would take to transition to a Net Zero global economy (IEA, 2017, 2021). This transition is not just about greenhouse gas emissions and the climate, it is also about nature and biodiversity (MEA, 2003; TEEB, 2010; SEEA, 2014; TCFD, 2017; CBD, 2021; IPBES, 2019; TNFD, 2021, Dasgupta, 2021).

At COP26 in Glasgow, in November 2021,[2] governments deliberated commitments and pathways to Net Zero, to achieving the Paris Agreement targets. While initiatives and announcements have been many, reactions have been mixed. While urgency, targets, tools, and frameworks remain the subject of debate, the recognition that the finance industry must rapidly adjust and adapt to this new reality is clear and unanimous. Indeed, UK's Chancellor of the Exchequer pledged to 'rewire' the global financial system for Net Zero (Herman, 2021).

Witnessing this growing awareness is truly encouraging. However, to devise the right solutions, to truly rewire the global financial system, we must understand how we ended up here, and why we have tolerated such levels of pollution and waste in our air, rivers, oceans, land, and even outer space. One of the main reasons for such a suboptimal outcome can be found in our financial value framework, in the principles of finance that have governed financial education and training, our markets and investments. The root cause is in the value equations of core finance theory and practice, in the equations that have been, and still are, focused on risk and time, serving one stakeholder, the risk-averse return maximising investor (Brealey et al., 2020; Choudhry, 2012; Damodaran, 2017; Friedman, 1970; Graham & Harvey, 2002; Koller et al., 2011; Markowitz, 1952).

Indeed, in theory and practice, finance scholars and investors have spent decades researching and debating whether or not it is possible to achieve abnormal risk-adjusted returns in the stock market—many still do. The Efficient Market Hypothesis (EMH), the many thousands of studies looking for evidence of stock market anomalies, they were all, ultimately, focused on finding out if an individual investor could achieve market-adjusted abnormal returns above the level of risk involved in an investment (De Bondt & Thaler, 1985; Dissanaike, 1997; Fama, 1970; Fama & French, 1992, 1993; Malkiel, 1973).

Interestingly, the space impact of the corporations whose stocks were being used to find statistical evidence of efficiency or random walks has not been

[2] The UK hosted the 26th UN Climate Change Conference of the Parties (COP26) in Glasgow between 31st of October and 13th of November 2021.

part of the discussion. Focused on abnormal risk-adjusted returns, the literature has not considered the impact-adjusted returns of the stocks included in the portfolios. Indeed, we all remember the many experiments in academia and industry trying to prove that a blindfolded dart throwing monkey could perform just as well as, if not better than, an expert fund manager.[3] It seems, experts were also blindfolded, by a framework of value focused on risk and time, a bias predicated by the logic of the value of money taught and built in the finance discipline.

Given the transformed landscape and the raging climate crisis, the finance industry is now actively developing a new framework through which it will align investment portfolios with our sustainability and climate targets. The same industry that brought the global financial system to collapse in 2008, that serves and has always served the individual risk-averse investor, disregarding impact and responsibility, is now self-regulating through a framework focused on the risks, opportunities, and financial impact of climate change—this is the purpose of the proposed framework for climate-related disclosures by the Financial Stability Board—Task Force on Climate-related Financial Disclosures (TCFD, 2017)—a framework that enjoys the support of more than 2600 companies in 89 countries, including 1069 financial institutions responsible for $194 trillion of assets (FSB, 2021).

The portfolio alignment tools and targets that have been put forward by and through the TCFD framework, although well-documented and well-intentioned, do not go far enough to trigger the deep and immediate change necessary to reverse the immense damage we have inflicted, and continue to inflict, on our home planet. Besides the fact that the framework is still a voluntary one,[4] its philosophy is that of engagement, not divestment. In other words, however high the implied temperature score of an asset or portfolio, the framework does not recommend, let alone require, divestment (TCFD-PAT, 2021).

While understandable from a systemic perspective and given the infancy of the framework, this approach poses a challenge to our pathways and Net Zero strategies. When an investment causes significant losses to the risk-averse investor and has very high risks of further losses, the finance industry would recommend to 'not invest' or 'divest', or maybe even 'short the asset'. Engagement is used only in some specific cases by specific types of investors and types of investment.

---

[3] These experiments tested Malkiel's (1973) claim and were widely reported and discussed in the media (Ferri 2012; Kueppers 2001).

[4] Soon to be mandatory in the UK (UK Government, 2021).

However, when an asset or investment is causing immense damage to the environment, has very high risks of further damage and a very high implied temperature score, the framework recommends engagement in all cases, even for those in the fossil fuel industry. If the framework were to be balanced, we should at least expect the same logic applied to investments that have a significant negative impact on the planet and humanity's well-being.

We are working with a framework that has variable pathways, variable targets, variable scenarios, variable benchmarks, variable methodologies, and variable recommendations, all of which are left to the discretion and depend on the capabilities of the institutions measuring the alignment of their own portfolios. Given the intellectual and actual history of the discipline and the industry, this framework and its alignment tools risk becoming a sophisticated distraction.[5]

The transition to a Net Zero economy implies that we must intensively invest in alternative energy infrastructures, in transforming our value chains, in cleaning the air, oceans, rivers, landfills, etc., but also, we must stop investing in new high carbon and waste opportunities, and we must at some point divest from existing high carbon and waste companies/projects. The logic and tools with which we are to achieve this are loosely lost in the word 'engagement'. This is so because the framework proposes implied temperature scores for identifying assets and portfolios that need alignment, but there is no clearly described logic or structure to the 'engagement' that will achieve actual transition steps for those assets and portfolios. Indeed, we also have many disagreements regarding what is green and sustainable and what is not (EU, 2020, 2022).

While the TCFD framework and the climate-related disclosures initiative are important and required, for a consistent and global change across industry and business, for effective change, we need to rethink our financial value framework and associated equations. Such that, a pollution-averse planet and humanity are made equal stakeholders alongside the risk-averse return maximising investor, and our mathematics of monetary value and return reflect investors' responsibility of impact and our sustainability objectives.

Indeed, to address the evolutionary challenges we have created for ourselves, and climate change is only one of them, we need to redefine the value of money beyond risk and time, by integrating the space impact of cash flows and assets into the value equations of those cash flows and assets.

---

[5] Interestingly, the fossil fuel industry was the largest delegate group at COP26, see McGrath (2021). Also, banks are still some of the major investors in the fossil fuel industry. A recent report reveals that banks have allocated more than $1.5 trillion to the coal industry since 2019 (Ainger 2022; Urgewald 2022, 2021).

Neither alignment tools nor temperature scores achieve such an integration, they are parallel exercises and do not alter our value equations and principles. The proposed temperature scores are more like indicators of misalignment, rather than actual tools of transformation. The same applies to the commonly used, yet incomparable, opinion points called ESG ratings or scores. They do not integrate impact into our value models and equations.

During COP26, the International Financial Reporting Standards (IFRS) Foundation announced that it is joining forces with the Climate Disclosure Standards Board (CDSB) and the Value Reporting Foundation (VRF—which includes the Integrated Reporting Framework and the Sustainability Accounting Standards Board—SASB) to create the world's first International Sustainability Standards Board (ISSB) which aims 'to develop—in the public interest—a comprehensive global baseline of high-quality sustainability disclosure standards to meet investors' information needs' (IFRS, 2021; SASB, 2020a, b). These standards are critical and necessary, but they do not change the content of our value models, they do not change the value equations and principles being taught and applied across finance theory and practice, across markets and investments.

Also during COP26, we heard about the 450 banks, asset managers, insurance companies, and pension funds with $130 trillion assets under management who have now committed to the Glasgow Financial Alliance for Net Zero (UNFCCC, 2021)—an initiative led by Mark Carney, the former Governor of the Bank of England and the UN Envoy for Climate Action (Metcalf and Morales, 2021). To use science-based guidelines to reach Net Zero carbon emissions by mid-century is a commendable target, but with an engagement framework that is still voluntary, flexible, and variable on all key issues, with temperature scores that do not integrate impact into value equations, and more than half the market not reporting emissions yet, this announcement is exactly what it sounds like, a big number in a small step towards a distant target.

The Glasgow Financial Alliance for Net Zero is without a doubt a positive step. The scepticism that has been directed at this announcement is understandable given the way it has been communicated. The $130 trillion figure is not a commitment of cash or expenditure, but the total assets under management of those banks and asset managers, which also include mortgages by the trillions. The irony is inescapable, nothing to worry about, those who would kick you and your family out of your home for missed mortgage payments are saving your planet.[6]

[6] Indeed, foreclosures have already started to rise since pandemic-related constraints have been removed. See Olick (2021).

Interestingly, in December 2021, a month after the grand headlines of COP26, UK's Financial Conduct Authority (FCA) and Prudential Regulation Authority (PRA) fined three major banks for different failings. Standard Chartered was fined £46.5 million by the PRA (Bank of England, 2021) for failing to be open and cooperative and for failings in its regulatory reporting governance and controls. HSBC was fined £63.9 million for deficient transaction monitoring controls (FCA, 2021b), NATWEST was fined £264.8 million for anti-money laundering failures (FCA, 2021a). In truth, the year 2021 was a record year for FCA fines. In October 2021, the FCA fined Credit Suisse (FCA, 2021c) over £147 million for 'serious financial crime due diligence failings related to loans worth over $1.3 billion'. In July 2021, the FCA fined Lloyds Bank General Insurance Limited £90 million 'for failing to ensure that language contained within millions of home insurance renewals communications was clear, fair and not misleading' (FCA, 2021d).

If laws and legal requirements are not enough to encourage compliance and enforce our standards, one wonders how a voluntary framework would do. Leaving the future of our planet and the fate of our children at the mercy of the discretion of banks and other financial institutions may not be a very wise strategy. This is not to say that banks are to blame for our current predicament. While finance and our monetary architecture have significant responsibility to bear, our challenge is systemic, the responsibility is shared, and the solutions are in our own value paradigm.

The main proposition of this book is quite simple, if we truly want to rewire the financial system, we must change and reform our financial value framework. We must transform the logic of the value of money, and introduce equations where impact is integrated into value. Indeed, finance theory and practice, through the omission of space and our responsibility in it, through an entirely risk and time-based conceptualisation of money and its valuation, have been directly and indirectly complicit in causing the immense environmental and social crises we face today.

Since the early beginnings and for many decades, the value framework of the finance discipline has been built to serve the risk-averse investor, ignoring, or omitting, the human collective and the planet. Indeed, the finance discipline and industry have operated in a world without 'space', and without considering the impact of investments on and in space as an integral element of the value of investments. Space here refers to our physical context stretching from the world of atoms, inside matter, to the planet, its core, surface, atmosphere, and outer space (I discuss and explore the many layers of space in more detail in chapter four).

This has been so because the analytical value framework of the discipline has been built around two key principles of value, (1) Risk and Return, and (2) Time Value of Money. Furthermore, financial models have been geared towards analysing the profitability of cash flows, rather than their creative impact, and the profitability of cash flows has been assessed based on the time and risk value of cash flows focused on investor returns without integrating the impact of cash flows in the valuation of cash flows. As the discussion in this book will reveal, our current value framework leaves our evolutionary investments in a blind spot due to the biases introduced through these principles.

Indeed, we have a missing principle in finance, a principle that establishes the value of cash flows vis-à-vis space. To this purpose, I have proposed the introduction of a third principle into core finance theory and practice, the Space Value of Money. The space value of money principle complements time value of money and risk and return. It establishes our spatial responsibility and requires that a dollar ($1) invested in space has at the very least a dollar's ($1) worth of positive impact on space. It offers a number of associated metrics through which we can quantify/design space impact and integrate it in the value equations of assets and cash flows (Papazian, 2011, 2013a, b, 2015, 2017, 2020, 2021a, b). The space value of money could be the theoretical link between sustainability and finance, and the principle that makes finance inherently sustainable.

The waste we have left behind in our oceans, in the air, on land, and outer space is evidence of the necessity to reimagine human productivity, and finance theory and practice have a central role to play. To facilitate a radical change in the way we express and deploy human productivity on Earth, in space, and how we value and monitor that productivity, we need a new principle and a transformed framework. Money and the equations that govern its value, deployment, and creation are key to that process.

This book introduces space as a dimension of analysis and entrenches our responsibility for impact on and in space, our physical context, into the value framework and core equations of finance theory and practice. It offers an alternative approach to value design, measurement, and creation, discussing the theoretical, mathematical, institutional, technological, and data elements of the transformation. It proposes alternative solutions that can facilitate the deployment of sustainable finance algorithms in the future, and the funding of key evolutionary investments/challenges like the transition to Net Zero.

Our responsibility for space impact is simultaneously our challenge and our salvation, the key to a healthier planet, and the sustainable expansion of human productivity across time and space.

The book has nine chapters including this introduction. The next eight chapters are briefly summarised below.

2 Finance: A Value Paradigm and Equations without Space—Chapter 2 explores core finance theory and models and identifies the key gaps and cracks through which, to date, sustainability and responsibility have slipped through into omission and neglect. This chapter demonstrates that space as a dimension of reality and our physical context has been absent from the discipline, and the impact of cash flows and assets on space have not been accounted for in our models and equations used to measure returns and to value cash flows. A risk and time-focused value paradigm, serving the risk-averse return maximising mortal investor, is identified as an evolutionary bottleneck.

3 Sustainable Finance: Frameworks without Value Equations—Chapter 3 is a snapshot of sustainable finance, looking at carbon budgets and pathways, standards and frameworks, industry and market initiatives, and the key challenges facing the field. It also explores some of the popular metrics, covering ESG factor integration through ratings, and other portfolio alignment tools, like carbon intensity and implied temperature rise. The chapter reveals that the frameworks, metrics, and ratings used in the field do not actually integrate the impact of investments into the value equations of those investments. The chapter also identifies a market topography of attitudes towards sustainability as a potential challenge for further development and introduces several dimensions that are omitted from the growing focus on ESG.

4 The Missing Principle: The Space Value of Money—Chapter 4 introduces the missing principle of finance theory, the space value of money, and discusses its rationale and core proposition. Given the introduction of space into the analytical framework of finance, the chapter also reveals and discusses the key characteristics of two new stakeholders. In a discipline that recognises the risk, time, and space value of money, alongside the risk-averse mortal investor, we have a pollution-averse planet and an aspirational human society as stakeholders. These stakeholders require that the space impact of an investment is integrated into the value equations used in the discipline. The principle allows a structured approach to help the transition to a Net Zero economy.

5 Quantifying Space Impact—Chapter 5 builds upon the principle of the space value of money and explores the elements and measures of space impact. It identifies the necessity to map the space impact of cash flows and assets as a primary step, as industries operate in and affect many layers of space. It introduces several metrics that allow the quantification of space impact across the many aspects of concern to the new stakeholders, addressing planetary, human, and economic impact intensity measures, and introducing the main concepts of Net Space Value, Gross Space Value, and the Space Growth Rate.

6 Integrating Impact into Value—Chapter 6 builds on the principle and space impact measures and metrics presented in chapter four and five and provides the methodology and logic of integration with a sample of our value models and equations in finance. The space value of money principle requires space impacts to be positive and thus ensures that no negative space impact is ever allowed or absolved. The chapter describes a possible approach to the optimisation of positive impact.

7 The Algorithms of Sustainable Finance—Chapter 7 discusses the algorithms of sustainable finance, addressing key data-related challenges, as well as the elements, equations, and tools necessary to build functional and relevant algorithms that could incorporate sustainable finance into our technology-driven markets. The chapter explores how the space value of money principle and associated metrics can provide a real and tangible way forward for building sustainable finance algorithms that would allow us to transcend opinion points, or ratings, and build the needed tools for data-driven sustainable trading.

8 Sustainable Money Mechanics in Space—Chapter 8 builds on all the insights and propositions in the book and applies them to the topic of money creation, exploring both fiat debt-based money and cryptocurrencies. It addresses money mechanics in the context of a transformed value framework in finance, exploring the changes necessary to introduce space impact responsibility into our monetary architecture. If investors must earn their returns with a positive impact, then money creators, whether they are banks or central banks or cryptocurrency issuers, must also follow the same principle. The chapter proposes an alternative approach to money creation, Value Easing, based on the introduction of a new non-debt high space impact instrument, Public Capitalisation Notes. PCNs can help fund the many evolutionary investments/challenges we face, like the transition to Net Zero and outer space exploration.

9 Conclusion—Chapter 9 summarises the main insights and propositions and concludes the book.

References

Ainger, J. (2022). Coal is still raising trillions of dollars despite green shift. BNN Bloomberg. https://www-bnnbloomberg-ca.cdn.ampproject.org/c/s/www.bnnbloomberg.ca/coal-is-still-raising-trillions-of-dollars-despite-green-shift-1.1723066.amp.html. Accessed 15 February 2022.

Aleichem, S. (1949). Tevye's daughters: collected stories of Sholom Aleichem. Translated by Frances Butwin. Crown.

Bank of England. (2021). PRA fines Standard Chartered Bank £46,550,000 for failing to be open and cooperative with the PRA. Press Release. Bank of England. https://www.bankofengland.co.uk/news/2021/december/pra-final-notice-to-standard-chartered-bank-dated-20-december-2021. Accessed 02 February 2022.

Boffo, R., Marshall, C., & Patalano, R. (2020). ESG investing: Environmental pillar scoring and reporting. OECD Paris. https://www.oecd.org/finance/esg-investing-environmental-pillar-scoring-and-reporting.pdf. Accessed 02 February 2022.

Brealey, A. R., Myers, C. S., & Allen, F. (2020). *Principles of corporate finance* (13th ed.). McGraw Hill.

CBD. (2021). A new global framework for managing nature through 2030. UN Convention on Biological Diversity. https://www.cbd.int/article/draft-1-global-biodiversity-framework. Accessed 02 February 2022.

Choudhry, M. (2012). *The principles of banking*. Wiley.

CISL. (2021). Understanding the climate performance of investment funds. Part 2: A universal temperature score method. Cambridge Institute for Sustainability Leadership. https://www.cisl.cam.ac.uk/download-understanding-climate-performance-investment-funds-part-2. Accessed 02 February 2022.

Damodaran, A. (2017). *Damodaran on valuation* (2nd ed.). Wiley.

Dasgupta, P. (2021). The economics of biodiversity: The Dasgupta review. HM Treasury. https://assets.publishing.service.gov.uk/government/uploads/system/uploads/attachment_data/file/962785/The_Economics_of_Biodiversity_The_Dasgupta_Review_Full_Report.pdf. Accessed 02 March 2021.

De Bondt, W. F. M., & Thaler, R. (1985). Does the stock market overreact? *The Journal of Finance*, *40*(3), 893–805. https://doi.org/10.2307/2327804. Accessed 02 February 2021.

Dissanaike, G. (1997). Do stock market investors overreact? *Journal of Business Finance and Accounting*, *24*(1), 27–50. https://onlinelibrary.wiley.com/, https://doi.org/10.1111/1468-5957.00093. Accessed 02 February 2020.

EU. (2020). Sustainable finance taxonomy—Regulation (EU) 2020/852. European Commission. https://ec.europa.eu/info/law/sustainable-finance-taxonomy-regulation-eu-2020-852_en. Accessed 02 February 2022.

EU. (2022). EU taxonomy for sustainable activities. European Commission. https://ec.europa.eu/info/business-economy-euro/banking-and-finance/sustainable-finance/eu-taxonomy-sustainable-activities_en. Accessed 12 February 2022.

Fama, E. F. (1970). Efficient capital markets: A review of theory and empirical work. *The Journal of Finance, 25*, 383–417.

Fama, E. F., French, K. R. (1992). The cross-section of expected stock returns. *The Journal of Finance, 47*, 427–465. https://doi.org/10.2307/2329112. Accessed 02 February 2021.

Fama, E. F., French, K. R. (1993). Common risk factors in the returns on stocks and bonds. *The Journal of Financial Economics, 33*, 3–56. https://doi.org/10.1016/0304-405X(93)90023-5. Accessed 02 February 2021.

FCA. (2021a). NatWest fined £264.8 million for anti-money laundering failures. Press Releases. Financial Conduct Authority. https://www.fca.org.uk/news/press-releases/natwest-fined-264.8million-anti-money-laundering-failures. Accessed 02 Fenbruary 2022.

FCA. (2021b). FCA fines HSBC Bank plc £63.9 million for deficient transaction monitoring controls. Press Releases. Financial Conduct Authority. https://www.fca.org.uk/news/press-releases/fca-fines-hsbc-bank-plc-deficient-transaction-monitoring-controls. Accessed 02 February 2022.

FCA. (2021c). Credit Suisse fined £147,190,276 (US$200,664,504) and undertakes to the FCA to forgive US$200 million of Mozambican debt. Financial Conduct Authority. https://www.fca.org.uk/news/press-releases/credit-suisse-fined-ps147190276-us200664504-and-undertakes-fca-forgive-us200-million-mozambican-debt. Accessed 02 February 2022.

FCA. (2021d). FCA fines LBGI £90 million for failures in communications for home insurance renewals between 2009 and 2017. Financial conduct Authority. https://www.fca.org.uk/news/press-releases/fca-fines-lbgi-90-million-failures-communications-home-insurance-renewals-2009-2017. Accessed 02 February 2022.

Ferri, R. (2012). Any monkey can beat the market. Forbes. https://www.forbes.com/sites/rickferri/2012/12/20/any-monkey-can-beat-the-market/?sh=40520bd8630a. Accessed 02 February 2022.

Friedman, M. (1970). A Friedman doctrine—The social responsibility of business is to increase its profits. The New York Times. https://www.nytimes.com/1970/09/13/archives/a-friedman-doctrine-the-social-responsibility-of-business-is-to.html. Accessed 02 February 2022.

FSB. (2021). 2021 Status report: Task force on climate-related financial disclosures. Financial Stability Board. https://fsb.org/wp-content/uploads/P141021-1.pdf.

Graham, J., & Harvey, C. (2002). How CFOs make capital budgeting and capital structure decisions. *Journal of Applied Corporate Finance, 15*, 8–23.

Herman, Y. (2021). Britain's Sunak pledges to 'rewire' global finance system for net zero. Reuters News. https://www.reuters.com/business/cop/britains-sunak-pledges-rewire-global-finance-system-net-zero-2021-11-03/. Accessed 12 December 2021.

IEA. (2021). Net Zero by 2050: A roadmap for the global energy sector. International Energy Agency. https://iea.blob.core.windows.net/assets/deebef5d-0c34-4539-9d0c-10b13d840027/NetZeroby2050-ARoadmapfortheGlobalEnergySector_CORR.pdf. Accessed 02 February 2022.

IFRS. (2021). IFRS Foundation announces International Sustainability Standards Board. https://www.ifrs.org/news-and-events/news/2021/11/ifrs-foundation-announces-issb-consolidation-with-cdsb-vrf-publication-of-prototypes/. Accessed 02 February 2022.

IPBES. (2019). The global assessment report on biodiversity and ecosystem services. Intergovernmental Science-Policy Platform on Biodiversity and Ecosystem Services. https://ipbes.net/system/files/2021-06/2020%20IPBES%20GLOBAL%20REPORT%28FIRST%20PART%29_V3_SINGLE.pdf. Accessed 02 February 2021.

IPCC. (2013). Climate change 2013: The physical science basis. Summary for Policymakers. Intergovernmental Panel on Climate Change. https://www.ipcc.ch/site/assets/uploads/2018/03/WG1AR5_SummaryVolume_FINAL.pdf. Accessed 02 February 2021.

IPCC. (2018). Summary for policymakers. In Global warming of 1.5 °C. IPCC. Available at https://www.ipcc.ch/site/assets/uploads/sites/2/2018/07/SR15_SPM_High_Res.pdf. Accessed 12 December 2020.

IPCC. (2021). Climate change 2021: The physical science basis. https://www.ipcc.ch/report/ar6/wg1/downloads/report/IPCC_AR6_WGI_SPM_final.pdf. Accessed 02 February 2022.

IPCC. (2022). Climate change 2022: Impacts, adaptation and vulnerability. Summary for Policymakers. Intergovernmental Panel on Climate Change. https://report.ipcc.ch/ar6wg2/pdf/IPCC_AR6_WGII_SummaryForPolicymakers.pdf. Accessed 28 February 2022.

Keynes, J. M. (1936). The general theory of employment, interest and money. Macmillan.

Koller, T., Dobbs, R., Huyett, B., McKinsey and Company. (2011). Value: The four cornerstones of corporate finance (6th ed.). Wiley.

Kueppers, A. (2001). Blindfolded monkey beats humans with stock picks. *Wall Street Journal*. https://www.wsj.com/articles/SB991681622136214659. Accessed 02 February 2022.

Lewis, M. (2010). *The big short*. W. W. Norton & Company.

Lowenstein, R. (2010). *The end of wall street*. Penguin Books.

Lucchetti, A., & Burns, J. (2008). Moody's CEO warned profit push posed a risk to quality of ratings. *Wall Street Journal*. October 23 2008. https://www.wsj.com/articles/SB122471995644960797. Accessed 02 February 2022.

Malkiel, B. G. (1973). A random walk down wall street. W. W. Norton.

Markowitz, H. (1952). Portfolio selection. *The Journal of Finance*, *7*, 77–91. https://doi.org/10.2307/2975974. Accessed 02 February 2021.

McGrath, M. (2021). COP26: Fossil fuel industry has largest delegation at climate summit. https://www.bbc.co.uk/news/science-environment-59199484. Accessed 02 February 2022.

MEA. (2003). Ecosystems and human well-being: A framework for assessment. Millennium Ecosystem Assessment. Island Press. http://pdf.wri.org/ecosystems_human_wellbeing.pdf. Accessed 02 February 2021.

Metcalf, T., & Morales, A. (2021). Mark carney unveils $130 trillion in climate finance commitments. Bloomberg News. https://www.bloomberg.com/news/articles/2021-11-02/carney-s-climate-alliance-crests-130-trillion-as-pledges-soar. Accessed 01 January 2022.

Olick, D. (2021). Foreclosures are surging now that Covid mortgage bailouts are ending, but they're still at low levels. https://www.cnbc.com/2021/10/14/foreclosures-surge-67percent-as-covid-mortgage-bailouts-expire.html. Accessed 02 February 2022.

Papazian, A. V. (2011). A product that can save a system: Public capitalisation notes. In Collected Seminar Papers (2011–2012) (Vol. 1). Chair for Ethics and Financial Norms. Sorbonne University.

Papazian, A. V. (2013a). Space exploration and money mechanics: An evolutionary challenge. International Space Development Hub. https://papers.ssrn.com/sol3/papers.cfm?abstract_id=2388010 Accessed 06 June 2021.

Papazian, A. V. (2013b). Our financial imagination and the cosmos. Cambridge Judge Business School. University of Cambridge. https://www.jbs.cam.ac.uk/insight/2013b/our-financial-imagination-and-the-cosmos/. Accessed 12 December 2021.

Papazian, A. V. (2015). Value of money in spacetime. The International Banker. https://internationalbanker.com/finance/value-of-money-in-spacetime/. Accessed 02 February 2021.

Papazian, A. V. (2017). The space value of money. Review of Financial Markets. Chartered Institute for Securities and Investment, London. 12: 11–13. Available at: http://www.cisi.org/bookmark/web9/common/library/files/sironline/RFMJan17.pdf Accessed 16 December 2020.

Papazian, A. V. (2020). An algorithm for responsible prosperity: A new value paradigm. http://dx.doi.org/10.2139/ssrn.3633763. Accessed 01 February 2021.

Papazian, A. V. (2021a). Towards a general theory of climate finance. http://dx.doi.org/10.2139/ssrn.3797258. Accessed 02 June 2021.

Papazian, A. V. (2021b). Sustainable finance and the space value of money. Cambridge Judge Business School. University of Cambridge. https://www.jbs.cam.ac.uk/insight/2021/sustainable-finance-and-the-space-value-of-money/. Accessed 02 February 2022.

PRI. (2020). PRI annual report. Principles of Responsible Investment. https://www.unpri.org/download?ac=10948. Accessed 06 June 2021.

Ratsimiveh, K., Hubert, P., Lucas-Leclin, V., & Nicolas, E. (2020). ESG scores and beyond Part 1—Factor control: Isolating specific biases in ESG ratings. FTSE Russell. https://content.ftserussell.com/sites/default/files/esg_scores_and_beyond_part_1_final_v02.pdf. Accessed 02 February 2022.

SASB. (2020a). SASB implementation supplement: Greenhouse gas emissions and SASB standards. Sustainability Accounting Standards Board. https://www.sasb.org/wp-content/uploads/2020a/10/GHG-Emmissions-100520.pdf. Accessed 02 February 2022.

SASB. (2020b). SASB human capital bulletin. Sustainability Accounting Standards Board. https://www.sasb.org/wp-content/uploads/2020b/12/HumanCapitalBulletin-112320.pdf. Accessed 02 February 2022.

SEEA. (2014). System of environmental economic accounting 2012—Central framework. United Nations, European Commission, International Monetary Fund, The World Bank, Organisation of Economic Co-operation and Development, Food and Agriculture Organization of the United Nations. https://unstats.un.org/unsd/envaccounting/seeaRev/SEEA_CF_Final_en.pdf. Accessed 02 February 2021.

TCFD. (2017). Final report: Recommendations of the task force on climate-related financial disclosures. https://assets.bbhub.io/company/sites/60/2020/10/FINAL-2017-TCFD-Report-11052018.pdf. Accessed 02 February 2021.

TCFD-PAT. (2021). Measuring portfolio alignment: Technical considerations. https://www.tcfdhub.org/wp-content/uploads/2021/10/PAT_Measuring_Portfolio_Alignment_Technical_Considerations.pdf. Accessed 12 December 2021.

TEEB. (2010). The economics of ecosystems and biodiversity: Mainstreaming the economics of nature: A synthesis of the approach, conclusions and recommendations of TEEB. http://www.teebweb.org/wp-content/uploads/Study%20and%20Reports/Reports/Synthesis%20report/TEEB%20Synthesis%20Report%202010.pdf. Accessed 02 February 2021.

TNFD. (2021b). Nature in scope: A summary of the proposed scope, governance, work plan, communication and resourcing plan of the TNFD. Taskforce on Nature-related Financial Disclosures. https://tnfd.global/wp-content/uploads/2021/07/TNFD-Nature-in-Scope-2.pdf. Accessed 02 February 2022.

UK Government. (2021). UK to enshrine mandatory climate disclosures for largest companies in law. UK Government. https://www.gov.uk/government/news/uk-to-enshrine-mandatory-climate-disclosures-for-largest-companies-in-law. Accessed 02 February 2022.

UNFCCC. (2015). Paris agreement. United Nations Framework Convention on Climate Change. https://unfccc.int/sites/default/files/english_paris_agreement.pdf. Accessed 02 December 2020.

UNFCCC. (2021). COP 26 and the Glasgow Financial Alliance for Net Zero (GFANZ). https://racetozero.unfccc.int/wp-content/uploads/2021/04/GFANZ.pdf. Accessed 02 February 2022.

Urgewald. (2021). Groundbreaking research reveals the financiers of the coal industry, Urgewald. https://urgewald.org/en/medien/groundbreaking-research-reveals-financiers-coal-industry. Accessed 12 December 2021.

Urgewald. (2022). Who is still financing the global coal industry? Urgewald. https://www.coalexit.org/sites/default/files/download_public/GCEL.Finance.Research_urgewald_Media.Briefing_20220209%20%281%29.pdf. Accessed 22 February 2022.

US Congress. (2008). House hearing, 110 congress. Credit Rating Agencies and the Financial Crisis https://www.govinfo.gov/content/pkg/CHRG-110hhrg51103/html/CHRG-110hhrg51103.htm. Accessed 02 February 2022.

US DOJ. (2013). Department of Justice sues standard & poor's for fraud in rating mortgage-backed securities in the years leading up to the financial crisis. Department of Justice. https://www.justice.gov/opa/pr/department-justice-sues-standard-poor-s-fraud-rating-mortgage-backed-securities-years-leading. Accessed 02 February 2020.

US DOJ. (2016). Goldman Sachs agrees to pay more than $5 Billion in connection with its sale of residential mortgage-backed securities. US Department of Justice Press Release, https://www.justice.gov/opa/pr/goldman-sachs-agrees-pay-more-5-billion-connection-its-sale-residential-mortgage-backed. Accessed 02 February 2021.

2

Finance: A Value Paradigm and Equations Without Space

How ridiculous are the boundaries of mortals!

Lucius Annaeus Seneca, Naturales Quaestiones, 65AD

That's one small step for a man, one giant leap for mankind.

Neil Armstrong, Apollo 11, 1969

Thomas Kuhn, discussing the development of science and the paradigms that emerge within scientific fields, states that 'no natural history can be interpreted in the absence of at least some implicit body of intertwined theoretical and methodological belief that permits selection, evaluation, and criticism' (Kuhn, 1962, 16). He defines a scientific paradigm as 'universally recognized scientific achievements that for a time provide model problems and solutions to a community of practitioners' (Kuhn, 1962, viii).

As of today, the analytical framework of the finance discipline, in industry and academia, rests on two core principles: Risk and Return and Time Value of Money. This focus reveals a value paradigm built and defined by *Risk* and *Time* parameters, without any formal consideration of *Space* as an analytical dimension, nor any assessment of impact on space as our physical context. A review of finance literature in industry and academia supports this observation.

A survey of finance textbooks substantiates this claim, revealing what is taught and applied in the field. Brealey et al. (2020), a 13th edition core textbook in corporate finance, built on the wider academic literature, is a typical

A. V. Papazian, *The Space Value of Money*,
https://doi.org/10.1057/978-1-137-59489-1_2

example. Similarly, in Pike et al. (2018), a 9th edition textbook on corporate finance and investment, and Watson and Head (2016), a 7th edition principles and practice handbook for corporate finance, we observe the same framework and principles at work.

In the professional banking and finance literature (Choudhry, 2012, 2018), we can see evidence of the same. In investment valuation (Damodaran, 2012, 2017), and company valuation (Koller et al., 2011, 2015), we encounter the same fundamental principles in action. In project finance (Yescombe, 2014), in investment banking (Rosenbaum & Pearl, 2013), in property valuation and investment (Isaac & O'Leary, 2013), we find the direct and indirect reference as well as use of these two principles across the board.

The references mentioned here are given as a small relatively recent sample selection of a much wider literature with diverse levels of complexity built around the two main principles of value. Indeed, a closer scrutiny of a selection of finance models in this chapter will further support this observation, revealing a value paradigm without space, without responsibility. Our financial value framework and the resulting equations used to value investments are, in truth, built in a risktime universe.

Meanwhile, the growing climate emergency and the undeniable evidence of human responsibility (IPCC, 2018, 2021, 2022) have triggered an entirely new chapter in finance and industry. As the debate on the standards, principles, tools, and equations of sustainable finance continues and intensifies (discussed in chapter three), we are still teaching and applying principles of finance that cater to a risktime universe, without any reference to space, without considering and integrating the space impact of cash flows and assets into the value/return equations of those cash flows and assets.

In the following sections, I discuss a selection of theoretical and mathematical expressions of this risk and time focused value framework that has dominated finance theory and practice for many decades.

2.1 The Risk and Time Value of Money

In order to understand the risk and time focus of the finance discipline, we must explore and understand the stakeholder that the field has served and continues to serve, both in theory and in practice. A focus on risk and time in finance is the result of an internal rationale that serves one key stakeholder, the risk-averse return maximising investor. While the theoretical reasons as to why this has been so can be debated,[1] much of its more recent applications can be linked to utility maximisation and rationality in neoliberal economics. Simon (2001) describes utility maximisation, and its universal application, as follows:

> Utility maximization, the best developed formal theory of rationality, which forms the core of neoclassical economics, does not refer to the social context of action. It postulates a utility function, which measures the degree to which an individual's (aggregate) goals are achieved as a result of their actions. The rational actor chooses the action, from among those given, which maximizes utility. If the actor's goals are food and sleep, then rationality calls for choosing the attainable combination of food and sleep that yields the greatest utility. The core theory does not specify the content of the utility function (Simon, 1983).

The risk-averse return maximising investor has been a central organising principle of the finance discipline, the prism through which corporate finance and financial management challenges are treated, considered relevant and in tune with the scientific body of knowledge. If at all, the environment and wider society have been treated as a qualitative addendum, a side discussion to the main mathematical concepts, usually treated under the broad umbrella of 'corporate social responsibility'. A treatment that is similar to how externalities have been explored in economics.

Indeed, the risk-averse *mortal* investor has been the direct and indirect beneficiary of many millions of dollars of research funding over the last many decades.[2] In the literature, the term does not include the word 'mortal', I have added the term in order to further clarify the mindset of this stakeholder. If the stakeholder in finance, the investor, were to be eternal or immortal, she/he/they would have a very different set of priorities. While it would be hard to state with any degree of certitude what the priorities of investors would be if investors were immortal, we could comfortably state that short-term defeats and/or failures would not be as important, and the investor

---

[1] The philosophical roots of this focus can potentially be traced back to moral philosophy and the utilitarianism of Jeremy Bentham and John Stuart Mill. Although with much reductionism and simplification, Adam Smith's invisible hand and the assumptions of classical economics along with Alfred Marshall's introduction of the utility function into economics have contributed to this focus. Friedman's doctrine (Friedman 1970) has provided the business rationale of shareholder value maximisation which is a key aspect of this focus.

[2] Note that the shareholder value maximisation rationale has also been critiqued extensively in the literature (See Stout, 2012, for a discussion of the theory, history, and critique of 'shareholder value maximisation').

would not be as risk averse and would be less concerned with time. After all, time and risk are very mortal concerns.[3]

Since Haynes (1895) and Knight (1921) risk and uncertainty have become central topics in economics and subsequently in finance. Haynes (1895) writes:

> The word risk has acquired no technical meaning in economics, but signifies here as elsewhere chance of damage or loss. The fortuitous element is the distinguishing characteristic of a risk. If there is any uncertainty whether or not the performance of a given act will produce a harmful result, the performance of that act is the assumption of a risk (Haynes, 1895, 409)

Knight (1921) introduced what has come to be known as Knightian uncertainty, distinguishing it from risk. Knight's distinction rests on defining risk as quantifiable, with probabilities that can be known, and uncertainty as unquantifiable, where probabilities are unknowable. In a much-quoted paragraph, he writes:

> Uncertainty must be taken in a sense radically distinct from the familiar notion of Risk, from which it has never been properly separated...The essential fact is that "risk" means in some cases a quantity susceptible of measurement, while at other times it is something distinctly not of this character; and there are far-reaching and crucial differences in the bearings of the phenomena depending on which of the two is really present and operating...It will appear that a *measurable* uncertainty, or "risk" proper, as we shall use the term, is so far different from an *unmeasurable* one that it is not in effect an uncertainty at all. We shall accordingly restrict the term "uncertainty" to cases of the non-quantitative type. It is this "true" uncertainty, and not risk, as has been argued, which forms the basis of a valid theory of profit and accounts for the divergence between actual and theoretical competition. (Knight, 1921, 19-20).

While Knight's proposition is interesting to consider, it has been the subject of intense debate, due mainly to the distinction between 'measurable' and 'unmeasurable' uncertainty. One of the more recent and outspoken critiques is Taleb (2007). Taleb argues that '[i]n real life you do not know the odds; you need to discover them'. Writing about Knight, he continues, '[h]ad he taken financial or economic risk he would have realised that these "computable"

[3] I would like to state clearly that there are many forward-looking risk-taking and legacy defining investors who invest to build and improve human society and the future of our children. The discussion here is about the value framework in finance theory and practice, and not a blanket qualification of actual investors.

risks are largely absent from real life! They are laboratory contraptions!' (Taleb, 2007, 127–128).

We may or may not agree with Taleb's critique, the point here is to recognise that outside of controlled scenarios and predefined games of chance, all unknowns are unknowns, and the probabilities we project onto possible projected outcomes do not actually make the unknowns known or measured, they simply project a measured interpretation of possibilities without removing the uncertainty attached to the end outcome or result, which can always be an event outside of the projected alternatives and their probabilities. Indeed, whatever the complexity of a probabilistic measurement, it does not unravel uncertainty.

Since Markowitz's seminal work on portfolio selection (Markowitz, 1952), risk has been a central theme in finance. Risk is understood on two main levels, systematic risk and unsystematic risk. Systematic risk is inherent to the landscape, while unsystematic risk is specific to the asset or instrument being explored and can be reduced or eliminated through diversification. Markowitz introduced the idea that it was important to go beyond an asset's standalone risk and look at the relative riskiness of different assets in a portfolio context. This is critical for an investor focused on maximising returns and minimising risks, focused on stock returns and on building efficient portfolios. This is what has come to be known as modern portfolio theory.

Relative volatility over time became central to how portfolios are diversified, and how their risks are measured. Indeed, the entire discussion that Markowitz has put forward, and the discussion in the post-modern portfolio theory (Rom & Ferguson, 1993), revolves around the selection of assets for investment based purely on their risk/time/return characteristics. In the case of the modern portfolio theory, using variances, and in the case of the post-modern portfolio theory, using only downside risk, the standard deviation of negative returns.

Whether we are looking at portfolios or standalone assets, whether we are looking at equity or debt investments, private or public assets, primary or secondary transactions, the fact is, the value framework of core finance theory and practice is focused on risk and time, with the risk-averse mortal investor as sole stakeholder. This is achieved through the core principles used and applied when valuing cash flows and investments. The two principles are described below.

Time Value of Money: A dollar ($1) today is worth more than a dollar ($1) tomorrow—because a dollar today can earn interest/return by tomorrow and be more than a dollar by tomorrow.

Risk and Return: The higher the risk the higher the expected return—given the risk-averse nature of investors, higher risks imply higher expectations of reward.

It is quite clear from the above that space is neither mentioned nor included. The impact of one dollar ($1) is not considered, while its risk and time characteristics are well explored. The mathematical models commonly used in the field reflect this focus.

The purpose of the following discussion is to demonstrate that this focus on risk and time has translated into equations of value and return that reveal this omission of space as an analytical dimension and of impact on space as our physical context. In the next sections, I explore a selection of commonly used and referenced value and return models in finance. This is by no means an exhaustive list of models used in the field (Rubinstein, 2006), and the selection does not suggest that others are less important. Moreover, while the list is not exhaustive, the discussion is not comprehensive as it does not delve into the many assumptions and the many variations, evidence, or critiques of the models. My purpose is not to debate them, but to demonstrate that across the board, space and space impact are absent from the framework and the equations of value and return.

2.1.1 Discounting Cash Flows and the Net Present Value Model

When we translate the risk and time-based paradigm into mathematical expressions of value, the most basic and yet most fundamental equations we encounter in finance, widely used across industry and academia, are the cash flow discounting models. While the mathematics of finance uses discounting in almost all value models across different lines of application, it does not always reveal the origin of this key valuation method.

A relatively recent working paper by Goetzmann (2004) tracks the discounting method to Leonardo of Pisa or Fibonacci in 1202. Goetzmann finds evidence that in his Liber Abaci (Fibonacci, 1202) Fibonacci was the first to develop the present value analysis for comparing the economic value of alternative contractual cash flows.

> As before, Fibonacci explains how to construct a multi-period discount factor from the product of the reciprocals of the periodic growth rate of an investment, using the model developed from mercantile trips in which a percentage profit is realized at each city. In this problem, he explicitly quantifies the difference in the value of two contracts due to the timing of the cash flows alone. As

> such, this particular example marks the discovery of one of the most important tools in the mathematics of Finance – an analysis explicitly ranking different cash flow streams based upon their present value. (Goetzmann, 2004, 27)

Whatever the historical origins of the discounting method, today, the most central of them all is the Net Present Value equation (NPV). Indeed, Graham and Harvey (2002) reveal that Net Present Value (NPV) is one of the most frequently used capital budgeting techniques by CFOs, along with the internal rate of return (IRR), which is the discount rate that equalises NPV to zero.

NPV epitomises a universe where time and risk parameters define the value of an investment opportunity or of a series of expected future cash flows. It is the theoretical cornerstone of all Discounted Cash Flow (DCF) models in the field. In fact, this discussion is relevant to all discounted cash flow models including bond and annuity valuation models. In the NPV equation (Eq. 2.1), the purpose is to calculate the present value of future expected cash flows. The equation discounts future expected cash flows CF_t by r, the return on an investment or opportunity with the same level of risk, and subtracts the initial investment (II) from that value (Fig 2.1).

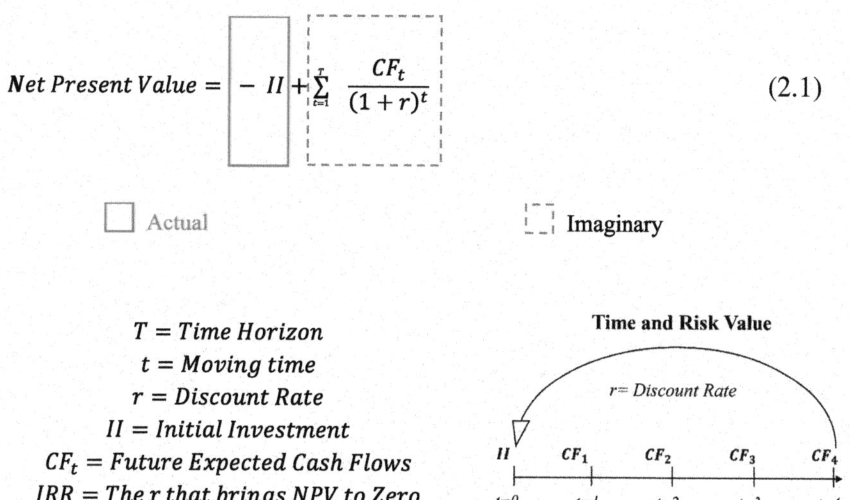

Fig. 2.1 The Net Present Value and cash flow timeline[4]

[4] The NPV equation is sometimes written in the below formats, where the first cash flow CF_0 (II) is included in the right-side term as the first cash flow at t = 0, or excluded but without the negative sign as the negative sign of the first cash flow CF_0 is assumed:

$$NPV = \sum_{t=0}^{T} \frac{CF_t}{(1+r)^t} \quad NPV = CF_0 + \sum_{t=1}^{T} \frac{CF_t}{(1+r)^t}$$

The logic of the model is straightforward. Given the time value of money, i.e., dollars in the future are worth less than dollars today because dollars today can earn an interest or return and be more than a dollar by tomorrow, expected or contractually agreed future cash flows must be discounted to the present to understand and gauge their current/present value. Given the risk and return relationship, the discount rate we use for this purpose is the return on an alternative investment with the same level of risk.

There are two parts to this equation, the actual part, which is what we would be investing to be able to expect the future expected cash flows, i.e., the initial investment (II), and the imaginary part or non-actual part in the future, the expected future cash flows. Future expected cash flows are imaginary because they have not happened yet. They may happen, or they may not, as expected or agreed. If these cash flows were guaranteed, it would not be necessary to factor in risk in the form of the discount rate. Worth noting that applying a discount rate to the future expected cash flows does not make the cash flows any less imaginary or more real.

The only certain element in the equation is the initial investment, and it is treated with a negative sign to indicate an outflow for the investor. Meanwhile, the imaginary part is mathematically treated for time and risk. It is quite surprising that the mathematical attention of the discipline has been entirely focused on the expected cash flows in the future, while the investment, that which is most certain in the present, is treated simply with a minus sign.

The omission of space happens when we ascribe an abstract negative sign to the initial investment, disregarding it entirely as a real process with a space impact. Indeed, the assessment of impact can be achieved by digging deeper into the many aspects of the investment and its deployment. In other words, the omission of space and space impact occurs when our treatment of the initial investment is defined and limited to ascribing a negative sign to denote an outflow for the risk-averse investor without further analysis or consideration of its utilisation and impact on the environment, on society, on space.

This is the theoretical and mathematical junction where we miss space, as an analytical dimension and our physical context. The above discussion is the proof of this omission.

2.1.2 Company and Stock Valuation Models

There are many different company and stock valuation models that seek to establish the fair value or intrinsic value of a stock or firm in the present.

The core principles and structure of these models are very similar to the Net Present Value model in the sense that the value of the asset or stock is defined based on the discounted present value of the future expected cash flows from the asset or stock. In the 6th Edition of their book *Valuation: Measuring and Managing the Value of Companies*, Koller et al. (2015) when discussing the 'Fundamental Principles of Value Creation' state and summarise how this discussion applies to companies:

> Companies create value for their owners by investing cash now to generate more cash in the future. The amount of value they create is the difference between cash inflows and the cost of the investments made, adjusted to reflect the fact that tomorrow's cash flows are worth less than today's because of the time value of money and the riskiness of future cash flows. (Koller et al., 2015, 17)

In the case of stocks, expected cash flows from the perspective of investors are defined by the Dividends (D) received. Which is why many stock valuation models focus on expected future dividends. Alternative models with a similar structure use Free Cash Flows to Firm (FCFF) or Free Cash Flows to Equity (FCFE) to value stocks, using the same logic and equations. Usually, FCFs are more relevant when Dividends are either not paid or much lower than the company's capacity to pay them.

Ultimately, whether using dividends or free cash flows, these models reflect the same principles of value discussed earlier, i.e., time value of money and risk and return, and following the previous discussion, omit space as a dimension and do not include investment impact as an integral part of the equations of value. Furthermore, value is once again understood as the discounted present value of future expected cash flows, whether in the form of dividends or free cash flows, and rests on a numerical adjustment to figures that are ultimately in the future, and therefore not actual. Meanwhile, the initial investment, i.e., the actual element, along with its space impact, are conveniently omitted.

The expected future in monetary terms receives more mathematical, as well as analytical, treatment than the actual impact that makes those expectations possible. I introduce and discuss a selection of these models in order to further demonstrate this point.

The Gordon model, as it has been commonly known and referred to, is one of the earliest, simplest, and yet most representative equations that describes the value of a stock. The model was developed by Williams (1938) and popularised by Gordon and Shapiro (1956).

While built upon many simplifying assumptions, and thus with many limitations, the model has been used to quantify the value of a stock, or its price, assuming dividends as the main expected cash flow from the stock, and a constant growth rate in those dividends. The time horizon applied is eternity, or infinity, as the corporation as a legal entity has no predetermined expiration date (Brealey et al., 2020).

Using the same principles, the equation for the intrinsic value of the stock is given as below:

$$P_0 = \sum_{t=1}^{\infty} \frac{D_t}{(1+r)^t} \tag{2.2}$$

Assuming a constant growth rate in future dividends, g, the model can be rewritten as:

$$P_0 = \frac{D_1}{r-g} \tag{2.3}$$

$$P_0 = Stock\ price$$

$$g = Constant\ Growth\ Rate\ in\ Dividends$$

$$r = Constant\ Cost\ of\ Capital$$

$$D_1 = Next\ Year/Period\ Dividend$$

The model uses dividends in the future, a constant growth rate in those dividends, and a cost of capital to reflect risk levels, without any reference to the impact of the corporation or the stock.[5]

Applying the same model to firm value, using Free Cash Flow to Firm or alternatively Free Cash Flow to Equity (FCFE), the model quantifying firm value is written as:

$$Firm\ Value = \sum_{t=1}^{\infty} \frac{FCFF_t}{(1+WACC)^t} \tag{2.4}$$

$$Firm\ Value = \frac{FCFF_1}{WACC-g} \tag{2.5}$$

[5] A look at 'Value Investing' built around earnings and fundamental value strategies discussed by Graham and Dodd (1934) in Security Analysis, and by Graham (1949) in The Intelligent Investor, reveals a similar abstraction from impact.

$$g = Constant\ Growth\ Rate\ in\ FCFF$$
$$WACC = Constant\ Cost\ of\ Capital$$
$$FCFF_1 = Next\ Year\ Free\ Cash\ Flow\ to\ Firm$$

Koller et al. argue that 'a company's return on invested capital (ROIC) and its revenue growth together determine how revenues are converted to cash flows (and earnings)' (2015, 17).

Adjusting to their specific ROIC and earnings growth proposition, they propose an alternative value driver formula based on the constant growth model:

$$Value = \frac{NOPLAT_{t=1} \times \left(1 - \frac{g}{ROIC}\right)}{WACC - g} \tag{2.6}$$

$$NOPLAT = Net\ Operating\ Profits\ Less\ Adjusted\ Taxes$$
$$g = Growth\ rate\ in\ NOPLAT$$
$$ROIC = Return\ on\ Invested\ Capital$$
$$WACC = Weighted\ Average\ Cost\ of\ Capital$$

Once again, none of these variables reflect or integrate the space impact of the company. In their multifaceted risk and time focused discussion of company value creation and management, Koller et al. (2015) do not discuss the space impact of cash flows. Space impact is not considered to be an element integral to the value of the cash flows and thus the companies under discussion.

Indeed, when alluding to climate change in the first introductory chapter, they state:

> For any company the complexity of addressing universal social issues like climate change poses an unresolved question: if the task does not fall to the individual company, then to whom does it fall? Some might argue that it would be better for the government to develop incentives, regulations, and taxes. In the example of climate change, this view might favor government action to encourage a migration away from polluting sources of energy. Others may espouse a free-market approach, allowing creative destruction to replace aging technologies and systems with cleaner and more efficient sources of power (Koller et al., 2015, 10).

By starting from the assumption/question 'if the task does not fall to the individual company, then to whom does it fall?' the authors have already externalised the responsibility of impact.[6]

Constant growth rates in dividends or in free cash flows are hardly a realistic assumption. They do, however, simplify the open-ended issue of what time horizon to use when valuing the future expected cash flows from a going concern. While there are many aspects that can be raised and debated, the key issue here is that none of the many versions and applications of the Gordon model involve the assessment of the space impact of investments. In other words, there is no reference to the space impact of the corporations involved, and there is no consideration of the specific space impact that generates those dividends and/or free cash flows.

Given that the constant growth assumption, though useful, is hardly ever an observed reality in the market, the model has been and is usually applied with a bit more sophistication. Unfortunately, this sophistication does not involve introducing impact measurement. It simply involves considering a variable growth rate in future dividends or future expected cash flows.

This is why the more commonly used stock valuation models are two-staged models, where for a specific limited period the growth rate in dividends is considered variable, usually a more manageable horizon, from five to ten years, and the remaining timeline is summarised in what is called a terminal value with a constant growth rate. Such that, the stock valuation equation becomes:

$$P_0 = \sum_{t=1}^{T} \frac{D_t}{(1 + WACC)^t} + \frac{P_T}{(1 + WACC)^T} \tag{2.7}$$

where P_T is the terminal value at time T, the last period of the variable rate time window, and summarises the value for the remaining future years where the growth rate is assumed to be constant.

$$P_T = \frac{D_{T+1}}{WACC - g} \tag{2.8}$$

[6] The authors have addressed the recent trend in sustainable finance by integrating ESG Factors in their 7th edition of the same book (2020). More on ESG factors in the next chapter.

Or alternatively,

$$P_0 = \sum_{t=1}^{T} \frac{D_t}{(1 + WACC)^t} + \frac{D_{T+1}}{(WACC - g) \times (1 + WACC)^T} \tag{2.9}$$

Applying the same equations to Free Cash Flows, and looking for firm value, we have:

$$Firm\ Value = \sum_{t=1}^{T} \frac{FCFF_t}{(1 + WACC)^t} + \frac{FCFF_{T+1}}{(WACC - g) \times (1 + WACC)^T} \tag{2.10}$$

In all of the above equations, WACC, or the Weighted Average Cost of Capital, is the cost of capital across all types of capital used, i.e., equity and debt, and when relevant can also include preferred stock. It is given by:

$$WACC = \left(\frac{E}{V} \times R_e\right) + \left[\left(\frac{D}{V} \times R_d\right) \times (1 - T)\right] \tag{2.11}$$

$$E = Market\ Value\ of\ Equity$$
$$D = Market\ Value\ of\ Debt$$
$$V = Total\ Value\ of\ Capital$$
$$\frac{E}{V} = \%\ of\ Capital\ in\ Equity$$
$$\frac{D}{V} = \%\ of\ Capital\ in\ Debt$$
$$R_e = Cost\ of\ Equity$$
$$T = Tax\ Rate$$

We observe that there is no reference to space impact. Time and risk are accounted for, and yet, no reference to the impact that it takes to achieve the Free Cash Flows. Indeed, space and impact are also omitted in the cost of capital calculations. This approach to valuation, where the impact of cash flows is externalised to the equations of value is exactly where finance absolves itself from responsibility.

In truth, if we were to think about these equations within the context of our current climate crisis, we can see why after decades of financial education and practice built around these principles and equations, we have reached this point of emergency.

The mathematical sophistication involved in these equations serves the individual risk-averse investor seeking to maximise her/his/their return while accounting for risk and time. There is no reference to responsibility for impact on planet and society, on space.

2.1.3 The CAPM: Capital Asset Pricing Model

The Capital Asset Pricing Model, developed by William Sharpe (1964) and John Lintner (1965), is another commonly used and applied model in theory and practice, with several variations and adjustments to its assumptions.

Describing the model's use and application, Fama and French (2004) write:

> The capital asset pricing model (CAPM) of William Sharpe (1964) and John Lintner (1965) marks the birth of asset pricing theory (resulting in a Nobel Prize for Sharpe in 1990). Four decades later, the CAPM is still widely used in applications, such as estimating the cost of capital for firms and evaluating the performance of managed portfolios. It is the centrepiece of MBA investment courses. Indeed, it is often the only asset pricing model taught in these courses (Fama & French, 2004, 25).

Since these papers, an immense amount of academic and technical research has been done to find evidence of correlation between Beta, the key risk parameter of the CAPM model, and expected returns on stocks (See Reinganum, 1981; Lakonishok & Shapiro, 1986; Fama & French, 1992; Gordon & Gordon, 1997). While the evidence has been contradictory and often non-existent, the model remains a commonly used theoretical tool.

The CAPM model states that the return on a security *i*, R_i, is equal to the risk-free rate, R_f, which is conceptually the return on an investment with zero risk and is usually considered to be the return on a government bond or T-Bill, plus a reward for the risk being taken by being in that asset or security, which is equal to the market premium, $R_m - R_f$, multiplied by a systemic risk volatility measure of the security *i*, β_i. The point here is that the return on a security or asset is measured by risk-based measures, which are in their turn measured by the time-based performance of returns.

$$R_i = R_f + \beta_i \times \left(R_m - R_f\right) \tag{2.12}$$

$$R_i = Return\ on\ security\ i$$
$$R_f = Risk\ Free\ Rate$$
$$\beta_i = Volatility\ measure$$
$$R_m = Return\ on\ market$$

As one can observe through the elements of the model, there is no reference to the impact of the asset or security on our planet, on the climate, on the environment, on society, etc. The logic of the model is based purely on risk/volatility.

Indeed, Beta, the risk proxy of the security, is measured through:

$$Beta_i = \beta_i = \frac{Covariance_{R_i, R_m}}{Variance_{R_m}} \tag{2.13}$$

where R_i is the return on the security *i*, and R_m is the return on the market. There are no climate or impact related considerations here, only a statistical measure that tracks the changes in the individual asset return and the market return, which, as many studies have shown, does not seem to correlate with expected returns (Reinganum, 1981; Coggin & Hunter, 1985; Lakonishok & Shapiro, 1986; Fama & French, 1992; Gordon & Gordon, 1997).

Indeed, our financial models seem to happen in a risktime universe, abstracted from space and the implications of our actions and investments on its many layers.

2.1.4 The Sharpe Ratio

Linked to the previous discussion, another relevant measure that is commonly used as a reference formula in finance and investment is the Sharpe Ratio, also developed by William Sharpe (1963, 1964), it aims to assess the return on risk by measuring the return above the risk-free rate achieved by a portfolio (R_p-R_f) by the standard deviation of the excess returns on the portfolio. The higher the ratio, the higher the risk-adjusted return of the portfolio.

$$Sharpe\ Ratio = \frac{R_p - R_f}{\sigma_p} \tag{2.14}$$

$$R_p = Return\ on\ Portfolio$$
$$R_f = Risk\ Free\ Rate$$
$$\sigma_p = Standard\ Deviation\ of\ the\ Portfolio's\ Excess\ Return$$

The returns and the statistical measure used in the equation are abstracted from impact. The impact of the stocks in the portfolio is not considered, just like it has been across the previous equations discussed in this chapter.

The portfolio is treated as a numerical or mathematical abstraction built on correlations between price/return data points, and statistical measures that gauge their divergence from the expected mean. The standard deviation of returns measures the amount of variation or dispersion of values from an expected value, or the mean of those returns. A high standard deviation would imply that observations are spread out over a wider value range.

Using a statistical concept, the standard deviation, as a volatility measure or proxy for riskiness, is common practice in finance. Interestingly, the measure of risk is the dispersion of returns rather than the impact on space. We observe an omission of space, and a focus on available risk and time related concepts and metrics.

2.1.5 The Arbitrage Pricing Theory and Three Factor Model

Ross (1971, 1976) is credited for developing the Arbitrage Pricing Theory (APT). The model is a multi-factor model where the asset returns are derived through a linear function of different factors and the assets' sensitivity to those factors. It is similar to the CAPM model but considers multiple factors and not just the market risk premium, and its assumptions are less restrictive.

This approach facilitates a more complex or more detailed integration of risk, by describing the return on an asset with respect to different factors. The three Factor Model developed by Fama and French (1993, 1996) is one such application where the additional risk taken by investors is conceived in relation to the market, size, and book to market factors.

> The model says that the expected return on a portfolio in excess of the risk-free rate [E(R_i) − R_f] is explained by the sensitivity of its return to three factors: (i) the excess return on a broad market portfolio (R_M − R_f); (ii) the difference between the return on a portfolio of small stocks and the return on a portfolio of large stocks (SMB, small minus big); and (iii) the difference between the return on a portfolio of high-book-to-market stocks and the return on a portfolio of low-book-to-market stocks (HML, high minus low). (Fama & French, 1996, 55)

$$E(R_i) - R_f = b_i\big(E(R_M) - R_f\big) + s_i E(SMB) + h_i E(HML) \quad (2.15)$$

where

$$E(R_i) - R_f = Expected\ Excess\ Return\ on\ Stock\ i$$
$$E(R_i) = Expected\ Return\ on\ Stock\ i$$
$$R_f = Risk\ Free\ Rate$$
$$E(R_M) = Expected\ Return\ on\ Market$$
$$E(R_M) - R_f = Expected\ Market\ Risk\ Premium$$
$$E(SMB) = Expected\ Size\ Premium$$
$$E(HML) = Expected\ Value\ Premium$$
$$b_i,\ s_i,\ h_i = Factor\ Sensititivies\ or\ Loadings$$

While interesting and informative in exploring past correlations between stock returns and specific factors, the APT and the Three Factor Model (and the five Factor Model also developed by Fama & French, 2015) do not define or include any impact consideration into the value of assets. Indeed, besides linking expected returns to a more complex equation of risk and other factors like size, book value, profitability, and investment, they do not provide any new principle or rationale of return.

2.1.6 Market Efficiency, Anomalies, and Risk-Adjusted Abnormal Returns

Following the analytical focus of the discipline, one of the most researched and written about topics in finance theory and practice is the subject of stock return predictability. This is conceptualised and understood as risk-adjusted abnormal returns because any return commensurate with the risk being undertaken is within the theoretical rationale of the risk and return principle.

Generation after generation, finance scholars and practitioners have dedicated their efforts to establishing the nature and inner workings of capital markets. More specifically, debating whether or not markets are efficient, whether or not there are market anomalies that can be used by investors to achieve risk-adjusted abnormal returns.

Indeed, we have spent decades debating and looking for statistical evidence of market efficiency, or market overreaction, mean reversion in returns, and correlations that could lead to investors achieving returns higher than what the risk levels of investments would imply. Since Eugene Fama's (1970) article that made the efficient market hypothesis a central thematic focus in the field, an enormous amount of research and analytical focus has been dedicated to finding out whether indeed stock prices reflect all publicly available information, thus eliminating any opportunities for abnormal returns soon after the opportunities arise.

Burton Malkiel (1973, 1992, 2003), building on Fama's work and his own, describes market efficiency in the New Palgrave Dictionary of Money and Finance as follows:

> A capital market is said to be efficient if it fully and correctly reflects all relevant information in determining security prices. Formally, the market is said to be efficient with respect to some information set, ϕ, if security prices would be unaffected by revealing that information to all participants. Moreover, efficiency with respect to an information set, ϕ, implies that it is impossible to make economic profits by trading on the basis of ϕ. (Malkiel, 1992)

Research on the weak form, semi-strong form, and strong form of market efficiency, using different markets, different information sets, etc., are numerous and offer very contradictory results.

Indeed, academic and industry research on stock market predictability is still ongoing. Xi et al. (2022) explore this literature in their recent article in the Journal of Finance. In their paper titled 'Anomalies and the Expected Market Return', they analyse the two main threads in the research. On the one hand, the literature that explores firm characteristics and their ability to predict cross-sectional returns (Fama & French, 1992, 1993, 1996, 2015; Harvey et al., 2016), and on the other, the literature that explores 'the *time-series* predictability of the aggregate market excess return based on a variety of economic and financial variables, such as valuation ratios, interest rates, and inflation' (Xi et al., 2022, 639). They look for linkages between these two approaches that have defined stock market predictability research for many decades.

The literature on this topic is simply too large to cover here. The relevant point to be made is that a vast segment of the literature has been looking for evidence, or lack thereof, that markets can be used to achieve abnormal returns, i.e., returns higher than warranted by the risk levels assumed by the risk-averse investor. The space impact of the corporations whose stocks are used to prove efficiency, overreaction, anomalies, and so on has not come into play in the statistical tests or the equations of return.

A special thread in the literature is the market overreaction literature (De Bondt & Thaler, 1985; Dissanaike, 1997), where the behavioural features of the risk-averse investor are integrated into the models, looking for overreaction to recent good and/or bad news, and thus significant stock price reversals.

When comparing or looking for price reversals, amongst other methodologies, the overreaction literature uses Rank Period Returns to organise and rank stocks and portfolios according to their past returns. The rank period returns (RPR) are calculated using:

$$RPR_{mr} = \prod_{t=-48}^{0} R_{it} - \prod_{t=-48}^{0} R_{mt} \tag{2.16}$$

where R_{it} is the return on security *i* in month *t*, and R_{mt} is the return on the market in month *t* using price relatives with dividends (Dissanaike, 1997). As noticed, the Rank Period Returns are based on the individual stock returns and the market returns, there is no formal consideration of impact when ranking past good performers and past bad performers. In a research paper shortlisted as a finalist during the 2004 competition for clinical papers in finance co-organised by the Journal of Financial Economics, Dissanaike and Papazian (2005) look into the firms included in the portfolios used to measure market overreaction, exploring a key aspect of corporate governance, i.e., management turnover. While the paper digs deeper by looking inside the firms, the focus is not on space impact.[7]

Market efficiency and market anomalies have been tested and proved or disproved according to risk and time patterns of returns. The space impact caused while achieving those returns has not been considered relevant. Indeed, until recently, the carbon footprint of companies listed on stock exchanges has not been considered 'material' information and has not been included in the many information sets used to test market efficiency.

The investor return in stock market investments is usually conceptualised as the sum of the Dividend Yield and the Capital Gains achieved in the period of holding the share. It is defined as follows:

$$Holding\ Period\ Return = HPR_t$$

$$HPR_t = \frac{DIV_t}{PPS_{t-1}} + \frac{PPS_t - PPS_{t-1}}{PPS_{t-1}} \tag{2.17}$$

[7] My own doctoral research (Papazian, 2004) utilised this RPR methodology to identify stock market winners and losers on the London Stock Exchange using the stocks included in the FT500 index between 1988 and 1992. I subsequently applied clinical methodology (Jensen et al., 1989) to explore and scrutinise the firms included in the portfolios in real time and space between 1988 and 1996.

$$DIV_t = Dividend\ at\ Time\ t$$
$$PPS_t = Price\ of\ Share\ or\ Stock\ at\ Time\ t$$
$$PPS_{t-1} = Price\ of\ Share\ or\ Stock\ at\ Time\ t-1$$

While studies exploring stock return predictability have used and debated a variety of ways to calculate the return to an investor (Dissanaike, 1994, 1997; Gordon & Gordon, 1997), the above equation describes what an investor would earn if she/he/they bought the share at time t − 1, and held it until time t. It does not include transaction costs and assumes that Dividends, if and when they occur in the holding period, accrue to the investor in question. The focus is on variables that do not involve any consideration for the impact of the corporation whose stock is being traded for return. The perspective is that of the risk-averse investor, and the space impact of the corporation behind the share price is not factored in.

2.1.7 The Modigliani Miller Theorem and Corporate Investment

Developed in the late 1950s and further improved in the 1960s by Franco Modigliani and Merton Miller, the MM theorem has been a central theoretical and strategic reference in corporate finance (Modigliani & Miller, 1958, 1963). Its main proposition states that in perfect capital markets, 'the market value of any firm is independent of its capital structure and is given by capitalising its expected return at the rate ρ_k appropriate to its class' (Modigliani & Miller, 1958, 268).

$$V_j = \left(S_j + D_j\right) = \frac{\overline{X_j}}{\rho_k} \tag{2.18}$$

This is built on the assumption that in any given class the price of every share must be proportional to its expected return, denoted by the factor of proportionality for any class k, by $1/\rho_k$.

$$V_j = Value\ of\ Firm\ j$$
$$S_j = Market\ Value\ of\ Common\ Shares\ of\ j$$
$$D_j = Market\ Value\ of\ Debts\ of\ j$$
$$\overline{\overline{X}}_j = Expected\ Return\ on\ the\ Assets\ owened\ by\ the\ company$$
$$\rho_k = Capitalisation\ Rate\ for\ shares\ in\ class\ k$$

Proposition II of the MM theorem states that 'the expected rate of return or yield i of any company j belonging to the kth class is a linear function of leverage as follows':

$$i_j = \rho_k + (\rho_k - r)\frac{D_j}{S_j} \tag{2.19}$$

'That is, the expected yield of a share of stock is equal to the appropriate capitalization rate ρ_k for a pure equity stream in the class, plus a premium related to financial risk equal to the debt-to-equity ratio times the spread between ρ_k and r' (Modigliani & Miller, 1958, 271).

Do you see any variable in these equations that incorporates the space impact of the firm into its value? The parameters included are risk and return on different financing instruments, incorporating the market value of those instruments. The MM propositions state that the value of the firm is not affected by the capital structure of the firm, and we are left to wonder what an impact-adjusted MM theorem would look like.

The space impact is absent from the discussion, space, and our impact on it and in it are entirely exogenous to the theoretical and mathematical propositions.

2.1.8 The Black and Scholes Option Pricing Model

Although this book will not delve into derivative instruments, given that they derive their value from other underlying assets, and thus their impact is directly dependent on the impact of the underlying assets, I would like to make a quick reference to the famous option pricing model put forward by Black and Scholes (1973). The Black–Scholes option pricing formula is well known and well referenced in finance. Indeed, the authors along with Robert Merton (1973) earned a Nobel Prize for their work. The equation provides a value of a European option and is given by the below formula (Nobel Prize, 1997):

$$C = SN(d) - Le^{-rt}N\left(d - \sigma\sqrt{t}\right) \tag{2.20}$$

where d is defined as:

$$d = \frac{\ln\frac{S}{L} + \left(r + \frac{\sigma^2}{2}\right)t}{\sigma\sqrt{t}} \tag{2.21}$$

where

$$C = Value\,of\,Call\,Option$$
$$N_d = Normal\,Distribution\,Function$$
$$t = Time\,to\,Maturity$$
$$L = Exercise(Strike)Price\,of\,Option$$
$$\sigma = Standard\,Deviation\,of\,Return\,on\,Stock$$
$$r = Risk\,Free\,Interest\,Rate$$
$$S = Current\,Stock\,Price\,or\,Asset\,Price$$

Standard deviation, or variance, is used, for volatility, the current stock price, the option exercise price, the time to maturity, and, of course, the risk-free rate. We can observe the absence of any variable that measures or considers the space impact of the underlying asset, or the space impact of the corporation underlying the underlying asset, in this case the stock.

2.1.9 Technical Analysis & Comparables

While many investors and asset managers seek to find and capitalise on value, there is also a significant segment in the market that focuses on market actions, analysing prices and volume through charts (Rhea, 1932; Edwards & Magee, 1948; Murphy, 1999; Credit Suisse, 2010).

Technical analysts and investors analyse historical price and volume data and charts. The underlying assumption is that through their buy and sell activities investors already reveal the value of the stocks being traded. Thus, this second order by proxy investment 'valuation' methodology is even further removed from considering the space impact of corporations as an element of the investment decision.

Another method of 'valuing' stocks or corporations is by using comparables, by benchmarking specific data points that position the corporation in question vis-à-vis others in the industry or market. This method, while not using charts, prices, and volume, uses other market accessible data points like Price to Earnings Ratio (PE) and Book to Market Value Ratio (BMVR). This comparative approach is equally removed from integrating space and the space impact of corporations and their stocks, and it is another by proxy approach to relative value.

2.2 Space in Finance

There are only a couple of papers in economics and finance that do refer to space, or spacetime. However, they do not introduce space as an analytical dimension into core finance theory, and they do not transform our equations through the entrenchment of responsibility for space impact into our equations. They do refer to our galactic reality and the relativistic nature of space and time, and as such, they are worth mentioning here—specifically, Krugman's paper on interstellar trade (1978), and Gaarder Haug's theoretical discussion of relativity in the context of finance (2004). While commendable and inspiring from the perspective of addressing non-conventional aspects of trade and finance, neither actually addresses the introduction of space and our responsibility of impact into our value framework and core principles.

Another very recent development is the Spatial Finance Initiative. While this initiative is new and does not address space within our financial value models, it is relevant to our discussion. The Spatial Finance Initiative is part of the UK Centre for Greening Finance and Investment (CGFI), established by the Alan Turing Institute, Satellite Applications Catapult, and the Oxford Sustainable Finance Programme.

Spatial finance, as it has been defined, is the integration of geospatial data and analysis into finance theory and practice. The idea is summarised as follows:

> Spatial Finance, the integration of geospatial data and analysis in financial theory and practice, allows financial institutions to understand and manage risks, opportunities and impacts related to climate and the environment in a granular and actionable way. Advancements in geospatial technologies and data science are making it possible to collect asset-level information in a consistent, timely and scalable manner, increasing transparency for investors, policymakers, and civil society alike by aggregating insights at a company, sector or portfolio level from the physical asset level upwards (CGFI-SFI, 2021)

While this initiative is an important indicator of our changing times, it is more focused on the use of geospatial data and climate risks (Caldecott, 2019; SFPUO, 2018), rather than the transformation of our value paradigm.

Indeed, I believe this book provides a theoretical framework through which the use of geospatial data can be integrated with our financial value framework, analysis, and investment practices. However, as the discussion in this chapter has shown, and the following chapters will demonstrate, before we can effectively select data points, and independently of where those data points come from, i.e., satellite imagery or company disclosures, we must reframe and reform our value framework.

2.3 Externalities

Last but not least, this discussion would be incomplete without a reference to what have come to be known as 'externalities'. The economics literature on externalities has a long history. Externalities are defined as a cost or benefit that are generated or caused by an economic activity, productive or other, and incurred by an unrelated third party. Positive externalities are the benefits, negative externalities are the costs. Indeed, pollution, a key environmental impact, is considered a negative externality.

OECD (2021) defines externalities as follows:

> Externalities refers to situations when the effect of production or consumption of goods and services imposes costs or benefits on others which are not reflected in the prices charged for the goods and services being provided (OECD, 2021).

The broader history of the concept can be traced back to Marshall (1890), who introduced and discussed the concepts of internal and external economies:

> We may divide the economies arising from an increase in the scale of production of any kind of goods, into two classes -firstly, those dependent on the general development of the industry; and, secondly, those dependent on the resources of the individual houses of business engaged in it, on their organization and the efficiency of their management. We may call the former *external economies,* and the latter *internal economies* (Marshall, 1890, 266).

Pigou (1912), building on Marshall's work, introduced the idea of divergence of private and social net product, and argued for necessary interventions when such divergence rendered a disservice to persons other than the contracting parties.

> Thus, incidental uncharged disservices are rendered to the general public, in respect of resources invested in the running of motor cars that wear out the surface of the roads. Similar disservices attend the resources invested in the erection of buildings in crowded centres; for, such buildings, by contracting the air-space and the playing-room of the neighbourhood, tend to injure the health and efficiency of people living there.... It is plain that divergences between

> private and social net product of the kind just considered cannot, like divergences due to tenancy laws, be mitigated by a modification of the contractual relation between any two contracting parties, because the divergence arises out of a service or disservice rendered to persons other than the contracting parties. It is, however, possible for the State, if it so chooses, to remove the divergence in any field by "extraordinary encouragements" or "extraordinary restraints" upon investments in that field (Pigou, 1912, 163–4).

Coase (1960) after a critique of Pigou's approach restates the challenge from a total cost and opportunity cost perspective. When dealing with the actions of businesses that cause harm to others, he derives the following conclusion:

> Analysis in terms of divergencies between private and social products concentrates attention on particular deficiencies in the system and tends to nourish the belief that any measure which will remove the deficiency is necessarily desirable. It diverts attention from those other changes in the system which are inevitably associated with the corrective measure, changes which may well produce more harm than the original deficiency.... Economists who study problems of the firm habitually use an opportunity cost approach and compare the receipts obtained from a given combination of factors with alternative business arrangements. It would seem desirable to use a similar approach when dealing with questions of economic policy and to compare the total product yielded by alternative social arrangements (Coase, 1960, 42–43).

Given the laissez-faire classical background, most of the discussions on externalities have been situated within the market failure and state intervention literature. Some advocate taxes, subsidies, fines, and other measures that aim to compensate for or avert negative externalities, others argue against such interventions. Indeed, the literature exploring and classifying externalities is vast and explores the issue from a variety of perspectives. Just as an example, in a recent article Nguyen et al. (2016) discuss the quantification of environmental externalities (EIA, 1995) through alternative monetisation models, aiming at the internalisation of such costs in the price of products. They focus on electrical energy from renewable and non-renewable sources, and compare two approaches, a corrective tax equal to the externalities and a reduced value added tax rate on environmentally friendly products.

In his Nobel Prize lecture, Nordhaus (2018) describes climate change as the ultimate challenge of economics and defines it as follows:

> I begin with the fundamental problem posed by climate change —that it is a public good or externality. Such activities are ones whose costs or benefits spill outside the market and are not captured in market prices. These include

> positive spillovers like new knowledge and negative spillovers like pollution (Nordhaus, 2018).

While this is undoubtedly true given the conceptual and theoretical foundations of the field, the focus on market prices reflecting such costs reveals the bias. While quantifying externalities is a must, internalising them in prices does not actually address the issue of responsibility. Potentially, it would account for these heretofore unaccounted social costs, but it does not prevent negative impact.

The content of this book relates to what are known as externalities, however, I find the very use of the term a challenge. We must internalise 'externalities' before they are misnamed as externalities. Moreover, the approach that aims at 'treating externalities' does exactly that, it treats them after the fact, and does not transform the value paradigm that allows them.

2.4 Conclusion

To date, space as an analytical dimension and our physical context has been omitted from core finance theory and practice. The many layers of space within which our companies and industries operate have been abstracted away, leaving us with equations and tools that measure the time and risk value of cash flows, without any formal considerations of space impact. This approach to the value of money is directly and indirectly responsible for the current crises we face.

The logic of the value of money we have taught and applied, and still do, through our core finance curriculum, literature, and industry, is built in a universe with risk and time as its primary axes and the risk-averse return maximising mortal investor as its main and only stakeholder. Indeed, a quick look at the enormous literature looking for market anomalies and risk-adjusted abnormal returns is enough to understand the opportunity cost of this focus in the discipline.

The current crises we face are the symptoms of a financial value framework built in risktime, where space as an analytical dimension has been absent, and our responsibility for impact on space as our physical context has not been integrated into the equations of value and return used to assess investments.

Indeed, it is by serving the priorities of the mortal risk-averse return maximising investor, a tiny piece in the chain of humanity, rather than the entire chain or the planet hosting the chain, that finance theory and practice have been complicit in bringing about the challenges we face, to a situation where a climate catastrophe is more plausible than not. Somehow our financial value models have catered well to the needs of the mortal individual in the chain, but not to our collective needs and interests in spacetime.

A world facing a climate emergency, environmental degradation, pollution in rivers, seas, oceans, air, and outer space is the statistically significant consequence of a finance discipline operating in an abstracted risktime universe without space and our responsibility of impact in it and on it.

To integrate sustainability into our financial value framework, we must introduce space into the analysis. By establishing the necessity to introduce space into our theoretical and mathematical models, we ensure that finance theory and practice consider the risk, time, and space value of money.

Before expanding on the formal introduction of space into our financial value framework, the next chapter explores the current and evolving reality, trends, standards, and metrics in the sustainable finance field, also referred to as climate finance, responsible finance, impact finance, and ESG integration.

References

Black, F., & Scholes, M. (1973). The pricing of options and corporate liabilties. *The Journal of Political Economy, 81*, 637–654. https://www.jstor.org/stable/1831029. Accessed 02 February 2021.

Brealey, A. R., Myers, C. S., & Allen, F. (2020). *Principles of corporate finance* (13th ed.). McGraw Hill.

Caldecott, B. (2019). Viewpoint: Spatial finance has a key. IPE Magazine, https://www.ipe.com/viewpointspatial-finance-has-a-key-role-/10034269.article Accessed 02 February 2022.

Carney, M. (2015). Breaking the tragedy of the horizon—Climate change and financial stability. Bank of International Settlements FSB. Available at https://www.bis.org/review/r151009a.pdf. Accessed 26 December 2020.

CGFI-SFI. (2021). State and trends in spatial finance. Centre for Green Finance and Investment - Spatial Finance Initiative. https://www.cgfi.ac.uk/wp-content/uploads/2021/07/SpatialFinance_Report.pdf. Accessed 02 February 2022.

Choudhry, M. (2012). *The principles of banking*. Wiley.

Choudhry, M. (2018). *Past, present, and future principles of banking and finance*. Wiley.

Coase, R. H. (1960). The problem of social cost. *Journal of Law and Economics, 3*, 1–44. https://www.jstor.org/stable/724810. Accessed 02 February 2021.

Credit Suisse. (2010). Technical analysis—Explained. https://www.credit-suisse.com/pwp/pb/pb_research/technical_tutorial_de.pdf. Accessed 02 February 2022.

Damodaran, A. (2012). *Investment valuation* (3rd ed.). Wiley.

Damodaran, A. (2017). *Damodaran on valuation* (2nd ed.). Wiley.

De Bondt, W. F. M., & Thaler, R. (1985). Does the stock market overreact? *The Journal of Finance*, *40*(3), 893–805. https://doi.org/10.2307/2327804. Accessed 02 February 2021.

Dissanaike, G. (1994). On the computation of returns in tests of the stock market overreaction hypothesis. *Journal of Banking & Finance*, *18*(6), 1083–1094. https://doi.org/10.1016/0378-4266(94)00061-1. Accessed 02 February 2022.

Dissanaike, G. (1997). Do stock market investors overreact? *Journal of Business Finance and Accounting*, *24*(1), 27–50. https://doi.org/10.1111/1468-5957.00093. Accessed 02 February 2021.

Dissanaike, G., & Papazian, A. V. (2005). Management turnover in stock market winners and losers: A clinical investigation. European Corporate Governance Institute. ECGI. Finance Working Paper N° 61/2004. https://papers.ssrn.com/sol3/papers.cfm?abstract_id=628382. Accessed 02 February 2021.

Edwards, R. D., & Magee, J. (1948). Technical analysis of stock trends. Stock Trend Service.

EIA. (1995). Electricity generation and environmental externalities: Case studies. Energy Information Administration. https://www.nrc.gov/docs/ML1402/ML14029A023.pdf. Accessed 02 February 2022.

Fama, E. F. (1970). Efficient capital markets: A review of theory and empirical work. *The Journal of Finance*, *25*, 383–417. https://doi.org/10.2307/2325486. Accessed 02 February 2021.

Fama, E. F., & French, K. R. (1992). The cross-section of expected stock returns. *The Journal of Finance*, *47*, 427–465. https://doi.org/10.2307/2329112. Accessed 02 February 2021.

Fama, E. F., & French, K. R. (1996). Multifactor explanations of asset pricing anomalies. *The Journal of Finance*, *51*, 55–84. https://doi.org/10.1111/j.1540-6261.1996.tb05202.x. Accessed 02 February 2021.

Fama, E. F., & French, K. R. (2004). The capital asset pricing model: Theory and evidence. *Journal of Economic Perspectives*, *18*, 25–46. https://www.aeaweb.org/articles?id=10.1257/0895330042162430. Accessed 02 February 2021.

Fama, E. F., & French, K. R. (2015). A five-factor asset pricing model. *Journal of Financial Economics*, *116*, 1–22. https://doi.org/10.1016/j.jfineco.2014.10.010. Accessed 02 February 2021.

Fibonacci, Leonardo of Pisa (1202) Liber Abaci. Translated by Sigler, LE 2002, *Fibonacci's Liber Abaci*, Springer.

Friedman, M. (1970). A Friedman doctrine—The social responsibility of business is to increase its profits. The New York Times. https://www.nytimes.com/1970/09/13/archives/a-friedman-doctrine-the-social-responsibility-of-business-is-to.html. Accessed 02 February 2022.

Gaarder Haug, E. (2004). Space-time finance: The relativity theory's implications for mathematical finance. Wilmott Magazine. July 2004: 2–15.

Goetzmann, W. N. (2004). Fibonacci and the financial revolution. National Bureau of Economic Research. Working Paper 10352. http://www.nber.org/papers/w10352. Accessed 02 February 2022.

Gordon, J. R., & Gordon, M. J. (1997). The finite horizon expected return model. *Financial Analysts Journal, 53*, 52–61. https://doi.org/10.2469/faj.v53.n3.2084. Accessed 02 February 2021.

Gordon, M. J., & Shapiro, E. (1956). Capital equipment analysis: The required rate of profit. *Management Science, 3*, 102–110. https://www.jstor.org/stable/2627177. Accessed 02 February 2021.

Gordon, M. J. (1959). Dividends, earnings, and stock prices. *The Review of Economics and Statistics, 41*, 99–105. https://doi.org/10.2307/1927792. Accessed 02 February 2021.

Graham, B., & Dodd, D. (1934). *Security analysis*. McGraw Hill.

Graham, B. (1949). *The intelligent investor*. Harper & Brothers.

Graham, J., & Harvey, C. (2002). How CFOs make capital budgeting and capital structure decisions. *Journal of Applied Corporate Finance, 15*, 8–23. https://doi.org/10.1111/j.1745-6622.2002.tb00337.x. Accessed 02 February 2021.

Harvey, R., Liu, Y., & Zhu, H. (2016). ... and the cross-section of expected returns. *The Review of Financial Studies, 29*, 5–68. https://doi.org/10.1093/rfs/hhv059. Accessed 02 February 2021.

Haynes, J. (1895). Risk as an economic factor. *The Quarterly Journal of Economics, 9*(4), 409–449. https://doi.org/10.2307/1886012. Accessed 02 February 2022.

IPCC. (2018). Summary for policymakers. In Global warming of 1.5 °C. IPCC. Available at https://www.ipcc.ch/site/assets/uploads/sites/2/2018/07/SR15_SPM_High_Res.pdf. Accessed 12 December 2020.

IPCC. (2021). Climate change 2021: The physical science basis. https://www.ipcc.ch/report/ar6/wg1/downloads/report/IPCC_AR6_WGI_SPM_final.pdf. Accessed 02 February 2022.

IPCC. (2022). Climate change 2022: Impacts, adaptation and vulnerability. Summary for Policymakers. Intergovernmental Panel on Climate Change. https://report.ipcc.ch/ar6wg2/pdf/IPCC_AR6_WGII_SummaryForPolicymakers.pdf. Accessed 28 February 2022.

Isaac, D., & O'Leary, J. (2013). *Property valuation techniques* (3rd ed.). Palgrave Macmillan.

Jensen, M. C., Fama, E., Long, J., Ruback, R., Schwert, G. W., & Warner, J. B. (1989). Editorial: Clinical papers and their role in the development of financial economics. *Journal of Financial Economics, 24*, 3–6. https://www.sciencedirect.com/science/article/abs/pii/0304405X8990069X?via%3Dihub. Accessed 02 February 2021.

Knight, F. H. (1921). Risk, uncertainty and profit. Houghton Mifflin Company.

Koller, T., Dobbs, R., Huyett, B., & McKinsey and Company. (2011). Value: The four cornerstones of corporate finance (6th ed.). Wiley.

Koller, T., Goedhart, M., Wessels, D., & McKinsey and Company. (2015). Valuation: Measuring and managing the value of companies (6th ed.). Wiley.

Krugman, P. (1978). The theory of interstellar trade. *Economic Enquiry, 48*, 1119–1123.

Kuhn, T. (1962). *The structure of scientific revolutions*. University of Chicago Press.

Lakonishok & Shapiro. (1986). Systematic risk, total risk and size as determinants of stock market returns. *Journal of Banking & Finance, 10*, 115–132. https://doi.org/10.1016/0378-4266(86)90023-3. Accessed 02 February 2022.

Lintner, J. (1965). The valuation of risk assets and the selection of risky investments in stock portfolios and capital budgets. *The Review of Economics and Statistics, 47*, 13–37. https://doi.org/10.2307/1924119. Accessed 02 February 2021.

Malkiel, B. G. (1973). *A random walk down wall street*. W. W. Norton & Company.

Malkiel, B. G. (1992). Efficient market hypothesis. In P. Newman, M. Milgate, & J. Eatwell (Eds.), *New Palgrave dictionary of money and finance*. McMillan.

Malkiel, B. G. (2003). The efficient market hypothesis and its critics. *Journal of Economic Perspectives, 17*, 59–82. https://www.aeaweb.org/articles?id=10.1257/089533003321164958. Accessed 02 February 2021.

Markowitz, H. (1952). Portfolio selection. *The Journal of Finance, 7*, 77–91. https://doi.org/10.2307/2975974. Accessed 02 February 2021.

Marshall, A. (1890). *Principles of economics*. Macmillan.

Merton, R. (1973). An intertemporal capital asset pricing model. *Econometrica, 41*, 867–887. https://doi.org/10.2307/1913811. Accessed 02 February 2021.

Modigliani, F., & Miller, M. H. (1963). Corporate income taxes and the cost of capital: A correction. *The American Economic Review, 53*, 433–443. https://www.jstor.org/stable/1809167. Accessed 02 February 2021.

Modigliani, F., & Miller, M. H. (1958). The cost of capital, corporation finance and the theory of investment. *The American Economic Review, 48*, 261–297. https://www.jstor.org/stable/1809766. Accessed 02 February 2021.

Murphy, J. J. (1999). *Technical analysis of the financial markets: A comprehensive guide to trading methods and applications*. New York Institute of Finance.

Nguyen, T. L. T., Laratte, B., Guillaume, B., & Hua, A. (2016). Quantifying environmental externalities with a view to internalizing them in the price of products, using different monetization models. *Resources, Conservation and Recycling, 109*, 13–23. https://doi.org/10.1016/j.resconrec.2016.01.018. Accessed 02 February 2021.

Nobel Prize. (1997). For a new method to determine the value of derivatives. The Nobel Prize. Press Release. https://www.nobelprize.org/prizes/economic-sciences/1997/press-release/. Accessed 02 February 2022.

Nordhaus, W. D. (2018). Climate change: The ultimate challenge for economics. Nobel Prize Lecture. https://www.nobelprize.org/uploads/2018/10/nordhaus-lecture.pdf. Accessed 02 February 2022.

OECD. (2021). Externalities. OECD. https://stats.oecd.org/glossary/detail.asp?ID=3215. Accessed 02 February 2022.

Papazian, A. V. (2004). An endoscopy on stock market winners and losers. Unpublished PhD Dissertation. Cambridge University Judge Business School Library.

Pigou, A. C. (1912). *Wealth and welfare*. Macmillan and Co.

Pike, R., Neale, B., Akbar, S., & Linslley, P. (2018). *Corporate finance and investment* (9th ed.). Pearson.

Reinganum. (1981). Misspecification of capital asset pricing: Empirical anomalies based on earnings' yields and market values. *Journal of Financial Economics, 9*, 19–46. https://doi.org/10.1016/0304-405X(81)90019-2. Accessed 02 February 2021.

Rhea, R. (1932). The Dow theory. 12th Printing. Barron's.

Rom, B. M., & Ferguson, K. (1993). Post-modern portfolio theory comes of age. *Journal of Investing, 3*, 11–17. https://doi.org/10.3905/joi.2.4.27. Accessed 02 February 2021.

Roll, R., & Ross, S. A. (1980). An empirical investigation of the arbitrage pricing theory. *The Journal of Finance, 35*, 1073–1103.

Rosenbaum, J., & Pearl, J. (2013). *Investment banking*. Wiley.

Ross, S. A. (1976) The arbitrage theory of capital asset pricing. *Journal of Economic Theory, 13*, 341–360. https://doi.org/10.1016/0022-0531(76)90046-6. Accessed 02 February 2021.

Ross, S. A. (1978). The current status of the Capital Asset Pricing Model (CAPM). *Journal of Finance, 33*, 885–890. https://doi.org/10.2307/2326486. Accessed 02 February 2021.

Rubinstein, M. (2006). *A history of the theory of investments*. Wiley.

SFPUO. (2018). Climate Risk Analysis from space: Remote sensing, machine learning, and the future of measuring climate-related risk. Sustainable Finance Programme. https://www.smithschool.ox.ac.uk/research/sustainable-finance/publications/Remote-sensing-data-and-machine-learning-in-climate-risk-analysis.pdf. Accessed 02 February 2021.

Sharpe, W. F. (1963). A simplified model for portfolio analysis. *Management Science, 9*, 277–293. https://www.jstor.org/stable/2627407. Accessed 02 February 2021.

Sharpe, W. F. (1964). Capital asset prices: A theory of market equilibrium under conditions of risk. *Journal of Finance, 19*, 425–442. https://doi.org/10.1111/j.1540-6261.1964.tb02865.x. Accessed 02 February 2021.

Sharpe, W. F. (1975). Adjusting for risk in portfolio performance measurement. *Journal of Portfolio Management, 1*, 29–34. https://doi.org/10.3905/jpm.1975.408513. Accessed 02 February 2021.

Sharpe, W. F. (1994). The Sharpe ratio. *The Journal of Portfolio Management, 21*, 49–58. https://doi.org/10.3905/jpm.1994.409501. Accessed 02 February 2021.

Simon, H. A. (1983). *Reason in human affairs*. Stanford University Press.

Simon, H. A. (2001). Rationality in society. *International Encyclopaedia of the Social & Behavioral Sciences*, 12782–12786. https://doi.org/10.1016/B0-08-043076-7/01953-7. Accessed 02 February 2021.

Stout, L. A. (2012). *The shareholder value myth: How putting shareholders first harms investors, corporations, and the public*. Berrett-Koehler Publishers.

Taleb, N. N. (2007). *The Black Swan: The impact of the highly improbable*. Random House.

Watson, D., & Head, A. (2016). *Corporate Finance: Principles and Practice* (7th ed.). Pearson.

Williams, J. B. (1938). *The theory of investment value*. Harvard University Press.

Xi, D., Yan, L., Rapach, D. E., & Zhou, G. (2022). Anomalies and the expected market return. *The Journal of Finance*, *77*, 639–681. https://doi.org/10.1111/jofi.13099. Accessed 20 February 2022.

Yescombe, E. R. (2014). *Principles of project finance* (2nd ed.). Academic Press.

3

Sustainable Finance: Frameworks Without Value Equations

It is better to be vaguely right than exactly wrong.

Carveth Read, Logic: Deductive and Inductive, 1898

[T]hose of us who have looked to the self-interest of lending institutions to protect shareholders equity, myself especially, are in a state of shocked disbelief.

Alan Greenspan, Testimony Before House Committee on Oversight and Government Reform, 2008 See US Congress, 2008, for full testimony.

The discussion in the previous chapter demonstrated that space, as a dimension of analysis and our physical context, has been abstracted out of core finance principles and equations. While well equipped with equations and models that measure returns and value based on risk and time parameters, to date, core finance equations do not account for space and our responsibility of impact in it and on it. The purpose of this chapter is to explore the growing field of sustainable finance, to identify its main contributions, and establish its links and relevance, if any, to core finance principles and equations of value.

Sustainability gained mainstream attention after the 2015 Paris Agreement (PA) and the adoption of the UN Sustainable Development Goals (UNFCCC, 2015; UN, 2016). While the progress on both has been slow and an uphill battle (UN, 2020), it is an indisputable fact that 2015 has been an inflection point. The necessity to fix our industrial, business, and financial practices so that we can prevent the continuous warming of our planet is now front-page news every day.

A. V. Papazian, *The Space Value of Money*,
https://doi.org/10.1057/978-1-137-59489-1_3

The Net Zero and Race to Zero initiatives have redefined public discourse on national and international levels, in business, finance, and industry. The transformation of our energy systems, our industrial and non-industrial productive activities, our markets and investments, is now irreversibly on the global agenda (IEA, 2017, 2021; UNFCCC, 2020).

This worldwide recognition of human responsibility for climate change and the necessity for action has been achieved through the continuous work done by the Intergovernmental Panel on Climate Change (IPCC, 2022, 2021, 2018a, 2018b, 2013, 1988). In parallel, the worldwide recognition of human responsibility for our devastating impact on nature and biodiversity is owed to a variety of global institutions, amongst them is the Intergovernmental Science-Policy Platform for Biodiversity and Ecosystem Services (CBD, 2021; Dasgupta, 2021; IPBES, 2019; MEA, 2003).[1]

The acknowledgement of the necessity for action thrusted sustainability into the limelight and transformed sustainable finance from a niche sector to a global mainstream imperative. The finance industry is undoubtedly engaged with this global transition. During the last many years, we have witnessed the growth and development of sustainability standards (EFRAG, 2021; IFRS, 2021; IIRC, 2013, 2021; GRI, 2021; SASB, 2020a, 2020b, 2020c, 2021; SEEA, 2014; WRI & WBCSD, 2004, 2011) and frameworks for climate and nature-related financial disclosures that aim to support the alignment of the sector with the sustainability targets (TCFD, 2017, 2021a, 2021b, 2021c; TNFD, 2021a, 2021b, 2021c).

Responding to growing public demand and regulations, we have also witnessed the growth of science-based targets, ESG ratings and factor integration, portfolio alignment frameworks and metrics like temperature scores, carbon reporting guidelines, and a host of solutions aimed at helping and implementing this transition in business, industry, and finance (CCC, 2019; CDP, 2020; CDP-WWF, 2020; CISL, 2019, 2021a, 2021b; EU, 2018, 2020a, 2020b, 2022; EU-TEG, 2019; ILB et al., 2020; HM Government, 2019; MSCI, 2018; PRI, 2016, 2020; SEC, 2022; SEBTi, 2021, 2022; S&P Global, 2022a; TCFD-PAT, 2020, 2021; UBS-RI, 2022).

In parallel, banks, asset managers, and pension funds have initiated their own drive to support the transition through the Glasgow Financial Alliance for Net Zero (GFANZ) (UNFCCC, 2021), which is an alliance with $130 trillion assets under management, now committed to support this Net Zero

[1] Naturally, this is built upon the dedication and hard work of thousands of individual scientists and activists, public and private charities, non-governmental organisations, think tanks, and research organisations, over the last many decades.

target.[2] Moreover, extensive effort is being invested to quantify the amount of new funding needed to transform our energy systems, as well as the value chains of businesses and industries (HSBC-BCG, 2021; CPI, 2021; McKinsey GI, 2022).

The pathways, targets, standards, frameworks, metrics, tools, methodologies, and funding mechanisms of the transition to a sustainable Net Zero world economy are being shaped through an intense global debate. The aim is to achieve a successful transition, the integration of sustainability into the theory and practice of business, industry, and finance.

This chapter is a critical discussion of what has come to be known as sustainable finance, or impact finance, or ESG integration, or climate finance. The purpose is to contribute to the debate, and reveal that, as of now, our standards, frameworks, and metrics, besides being variable and voluntary, they do not go far enough to offer new value equations. Furthermore, they do not transform the existing equations of value and return discussed in chapter two. In other words, while sustainability and sustainable finance are moving to the mainstream and are ushering in changes in financial regulation, disclosure requirements, and business strategies, this rise to prominence is yet to penetrate the analytical content, framework, and equations of core finance theory and practice.

Naturally, given the size of the literature and the exponential increase in suggested approaches from a variety of sources, I address the key concepts and most prominent institutional frameworks, tools, and metrics.

3.1 Carbon Budgets and Pathways

The Paris Agreement set clear targets: to keep world temperature increases well below *2* °C above pre-industrial levels and to pursue efforts to limit the temperature increase to *1.5* °C above pre-industrial levels (UNFCCC, 2015). The Net Zero initiative, the transition to a low carbon economy, and the many steps being taken by governments and multilateral organisations are aimed at achieving this objective.

The systemic transformations necessary are serious and far-reaching, from energy systems to industrial practices, from business models to value chains, across sectors and countries, the leap required is nothing short of evolutionary.

---

[2] As this book was going to print, GFANZ released its guidance for a Net-zero Transition Plan, seeking inputs from the general public. Unfortunately, it has not been possible to include those plans in this analysis due to time constraints and production schedules.

One of the key concepts used to operationalise the transition and relevant strategies is the carbon budget. Carbon budgets help us define how much more emissions we can afford before overshooting the Paris Agreement temperature targets. They are helpful policy tools that allow relatively simplified assessments of our progress. In reality, and from a climate science perspective, there are many uncertainties and many moving parts to carbon budgets. Specifically, advances in carbon capture technology and their cost-effective application play a critical role in making these budgets relevant.

The IPCC (2018a) describes the carbon budget we have left to spend and identifies the probabilities associated with different budgets for achieving our targets.

> Limiting global warming requires limiting the total cumulative global anthropogenic emissions of CO_2 since the preindustrial period, that is, staying within a total carbon budget. By the end of 2017, anthropogenic CO_2 emissions since the pre-industrial period are estimated to have reduced the total carbon budget for 1.5 °C by approximately 2200 ± 320 $GtCO_2$. The associated remaining budget is being depleted by current emissions of 42 ± 3 $GtCO_2$ per year. The choice of the measure of global temperature affects the estimated remaining carbon budget. Using global mean surface air temperature gives an estimate of the remaining carbon budget of 580 $GtCO_2$ for a 50% probability of limiting warming to 1.5°C, and 420 $GtCO_2$ for a 66% probability. (IPCC, 2018a, 2018b, 12)

In a more recent report, IPCC (2021) has updated the residual carbon budget figures. The residual global carbon budget aimed at a 67% probability of remaining within the 1.5 °C target is given as 400 $GtCO_2$ from the beginning of 2020. Without significant reductions, with increasing world population and current global emission figures, we will consume the 400 $GtCO_2$ residual budget within a decade or so.

There are many institutions and initiatives that have been actively modelling alternative scenarios and pathways to a future that achieves our targets. Some aim and conceptualise meeting a 1.5 °C target, others 2 °C. These mitigation pathways, as they are known, are modelling exercises rather than actual policies. They are conceptual tools to enhance climate action on behalf of all involved parties, in order to fulfil the promise of the Paris Agreement.

The pathways are possible roadmaps that identify a combination of technological, industrial, economic, social, financial, and governance actions, policies, and deliverables as steps needed at different intervals of time in the future to achieve the vision by 2050. These pathways consider these variables and deliverables not just on a global scale, but also on a sectoral level. Thus, variations are many and the need to update them as we make more progress is an inbuilt necessity.

IPCC (2018a) states:

> Pathways consistent with 1.5 °C of warming above pre-industrial levels can be identified under a range of assumptions about economic growth, technology developments and lifestyles. However, lack of global cooperation, lack of governance of the required energy and land transformation, and increases in resource-intensive consumption are key impediments to achieving 1.5°C pathways. (IPCC, 2018a, 2018b, 95)

The Net Zero by 2050 pathway put forward by the International Energy Agency (IEA, 2021), see Table 3.1, focuses on the global energy sector and reveals fundamental milestones necessary to achieve an energy infrastructure that will sustain our climate ambitions beyond rhetoric. The energy sector is

Table 3.1 Key milestones to Net Zero by 2050 (Adapted from IEA, 2021)

	2021-2025	2030	2035	2040	2045-2050
Buildings	No new sales of fossil fuel boilers.	Universal energy access. All new buildings are zero-carbon-ready.	Most appliances and cooling systems sold are best in class.	50% of existing buildings retrofitted to zero-carbon-ready levels.	50% of heating demand met by heat pumps. More than 85% of buildings are zero-carbon-ready.
Transport		60%of global car sales are electric.	50% of heavy truck sales are electric. No new ICE car sales.	50% of fuels used in aviation are low emissions.	
Industry		Most new clean technologies in heavy industry demonstrated at scale	All industrial electric motor sales are best in class	Around 90% of existing capacity in heavy industries reaches end of investment cycle	More than 90% of heavy industrial production is low emissions
Electricity and Heat	No new unabated coal plants approved for development.	1020 GW annual solar and wind additions.	Overall net-zero emissions electricity in advanced economies.	Net-zero emissions electricity globally.	Almost 70% of electricity generation globally from solar PV and wind.
Other	No new oil and gas fields approved for development; no new coalmines or mine extensions.	Phase-out of unabated coal in advanced economies.		Phase-out of all unabated coal and oil power plants.	

the source of around three-quarters of greenhouse gas emissions today and is therefore key to any effective transition process.

Carbon budgets and mitigation pathways to Net Zero by 2050 are all attempts at finding the right course of policies, actions, and metrics through which we can actually change course, achieve the Paris targets, and save ourselves from irreversibly destroying our own ecosystem. Reducing emissions to Net Zero by 2050 is consistent with the target to limit the long-term increase in average global temperatures to 1.5 °C.

As the world strives to achieve this transition to Net Zero, alongside new energy systems, transformed industrial value chains, and revised business practices, the need to integrate and entrench sustainability in finance and investments has become a strategic priority.

In line with such an objective, we have witnessed the birth of new standards and frameworks, new tools and metrics, as well as a variety of approaches to their implementation. The following sections explore the most central developments.

As the discussion will show, the standards, frameworks, metrics, and tools available to date do not transform our core finance principles and equations. Moreover, they do not go far enough to offer new equations of value.

3.2 Sustainability: Standards and Frameworks

Over the last many years, a number of institutions and industry associations have initiated and led the development of sustainability accounting standards in order to facilitate the reporting of sustainability related data points. In parallel, a number of academics have actively advocated to engage the world's major accounting standard setters to address this opportunity (Barker and Eccles, 2018).

In 2021, during COP26, the International Financial Reporting Standards Foundation (IFRS, 2021) announced that it is joining forces with the Climate Disclosure Standards Board (CDSB) and the Value Reporting Foundation (VRF—which includes the Integrated Reporting Framework and the Sustainability Accounting Standards Board—SASB) to create the world's first International Sustainability Standards Board (ISSB) which aims 'to develop—in the public interest—a comprehensive global baseline of high-quality sustainability disclosure standards to meet investors' information needs' (IFRS, 2021).

Indeed, the history of these organisations started much earlier than 2021, some date back to the late 1990s, like the Global Reporting Initiative (GRI), others like the Sustainability Standards Board was created in 2011, and the Value Reporting Foundation in 2010. The European Sustainability Reporting Standards (ESRS) initiated by the European Financial Reporting Advisory Group is still in active development (EFRAG, 2021). The point here is to communicate to the reader that the efforts to build and develop a coherent and consistent set of sustainability standards are relatively new and are still in their early stages of standardisation. To put this in context, the Institute of Chartered Accountants in England and Wales (ICAEW) was established by royal charter in 1880, and the first 'Recommendations on Accounting Principles' in the UK was published in 1942 (ICAEW, 2021).

While the accounting standards for sustainability are important, and necessary, they do not change the content of our value models, and they do not change the value equations and principles being taught and applied across finance theory and practice, across markets and investments. The SASB standards (SASB, 2020a, b) download page describes the purpose and the functional relevance of the standards as follows:

> SASB Standards identify the subset of environmental, social, and governance issues most relevant to financial performance in each of 77 industries. They are designed to help companies disclose financially-material sustainability information to investors. (SASB, 2021)

While the standards themselves are not the subject of this section, and fall outside the scope of this book, the point here is quite straightforward. The disclosure of sustainability-related financially material information to investors, while central to the challenge, does not define nor change the value models and equations used by investors, or taught and applied in finance theory and practice.

The standards that define and structure sustainability-related material information for investors can organise and clarify the flow of information, but they do not interpret the information. Accounting data points have to be interpreted by investors to have any relevance to the financing and investing process. Indeed, they would make no sense without a framework of analysis and interpretation. Which brings us to the subject of frameworks within the sustainability field.

SASB (2022) discussing the growing number of standards and frameworks in the field and their relationship states the following:

> It is important to distinguish between sustainability frameworks and sustainability standards. Frameworks provide principles-based guidance on how information is structured, how it is prepared, and what broad topics are covered.

> Meanwhile, standards provide specific, detailed, and replicable requirements for what should be reported for each topic, including metrics. Standards make frameworks actionable, ensuring comparable, consistent, and reliable disclosure. Frameworks and standards are complementary and are designed to be used together (SASB, 2022)

Interestingly, neither standards nor frameworks change our value equations.

3.2.1 The TCFD Framework and Climate Disclosures

The TCFD[3] has become the main finance industry framework for climate-related disclosures, enjoying the support of more than 2,600 companies in 89 countries, including 1069 financial institutions responsible for assets of $194 trillion (FSB, 2021a; TCFD, 2017). In a 2019 report published by SASB and CDSB, the relationship between these frameworks/standards is defined as follows:

> The TCFD recommendations serve as a global foundation for effective climate-related disclosures. The CDSB Framework helps organizations integrate and disclose financially material climate and natural capital-related information into their annual reports. The SASB standards help organizations to collect, structure, and effectively disclose related performance data for the material, climate-related risks and opportunities they have identified (CDSB-SASB, 2019, 4)

The TCFD framework is still voluntary, although it will soon be mandatory in the UK (Bank of England, 2020). Based on the most recent figures published by MSCI, less than 40% of MSCI ACWI Investable Market Index constituents reported Scope 1 and 2 emissions, and less than 25% of MSCI ACWI IMI constituents reported Scope 3 GHG emissions (Bokern, 2022).

While the commitment to a roadmap that will entrench climate-related disclosures into financial reporting is being shaped as we speak, it is quite fascinating to note that the TCFD report describes the main analytical mindset and rationale as follows:

[3] The FSB-TCFD (Financial Stability Board—Task Force on Climate Related Financial Disclosures) framework, as it is in its name, is primarily focused on climate change and its purpose is 'to develop voluntary, consistent climate-related financial disclosures that would be useful to investors, lenders, and insurance underwriters in understanding material risks.... its members were selected by the Financial Stability Board and come from various organisations, including large banks, insurance companies, asset managers, pension funds, large non-financial companies, accounting and consulting firms, and credit rating agencies' (TCFD, 2017, iii).

> Those organizations in early stages of *evaluating the impact of climate change on their businesses and strategies* can begin by disclosing climate-related issues as they relate to governance, strategy, and risk management practices. The Task Force recognizes the challenges associated with measuring the impact of climate change but believes that by moving climate-related issues into mainstream annual financial filings, practices and techniques will evolve more rapidly (TCFD, 2017, v).[4]

It is quite clear from the text and the framework that we are, in truth, measuring the impact of climate change on the businesses and the financial system, and not the other way around. This is why the framework is identified as one of Climate-Related Risks, Opportunities, and Financial Impact.

The Final Report of Recommendations was published by the TCFD in June 2017 where a framework of four pillars and associated disclosure items were identified. The TCFD recommendations and suggested disclosures are given in Table 3.2 and Fig. 3.1.

The opportunities that the report discusses, as climate-related opportunities are defined as:

> Efforts to mitigate and adapt to climate change also produce opportunities for organizations, for example, through resource efficiency and cost savings, the adoption of low-emission energy sources, the development of new products and services, access to new markets, and building resilience along the supply chain (TCFD, 2017, 6).

It is evident from the model and the recommendations of the framework that before speaking of the responsibilities that must be assumed regarding climate change, we are focused on opportunities, risks, and their financial impact.

The report goes on to state:

> Better disclosure of the financial impacts of climate-related risks and opportunities on an organization is a key goal of the Task Force's work. In order to make more informed financial decisions, investors, lenders, and insurance underwriters need *to understand how climate-related risks and opportunities are likely to impact an organization's future financial position as reflected in its income statement, cash flow statement, and balance sheet* (TCFD, 2017, 8).[5]

---

[4] Emphasis added.

[5] Emphasis added.

Table 3.2 TCFD recommendations and suggested disclosures (TCFD, 2017)

Recommendations			
Governance	*Strategy*	*Risk management*	*Metrics and targets*
Disclose the organisation's governance around climate-related risks and opportunities	Disclose the actual and potential impacts of climate-related risks and opportunities on the organisation's businesses, strategy, and financial planning where such information is material	Disclose how the organisation identifies, assesses, and manages climate-related risks	Disclose the metrics and targets used to assess and manage relevant climate-related risks and opportunities where such information is material
Supporting Recommended Disclosures			
(a) Describe the board's oversight of climate-related risks and opportunities	(a) Describe the climate-related risks and opportunities the organisation has identified over the short, medium, and long term	(a) Describe the organisation's processes for identifying and assessing climate related risks	(a) Disclose the metrics used by the organisation to assess climate-related risks and opportunities in line with its strategy and risk management process
(b) Describe management's role in assessing and managing climate-related risks and opportunities	(b) Describe the impact of climate-related risks and opportunities on the organisation's businesses, strategy, and financial planning	(b) Describe the organisation's processes for managing climate-related risks	(b) Disclose Scope 1, Scope 2, and, if appropriate, Scope 3 greenhouse gas (GHG) emissions, and the related risks
	(c) Describe the resilience of the organisation's strategy, taking into consideration different climate-related scenarios, including a 2 °C or lower scenario	(c) Describe how processes for identifying, assessing, and managing climate related risks are integrated into the organisation's overall risk management	(c) Describe the targets used by the organisation to manage climate-related risks and opportunities and performance against targets

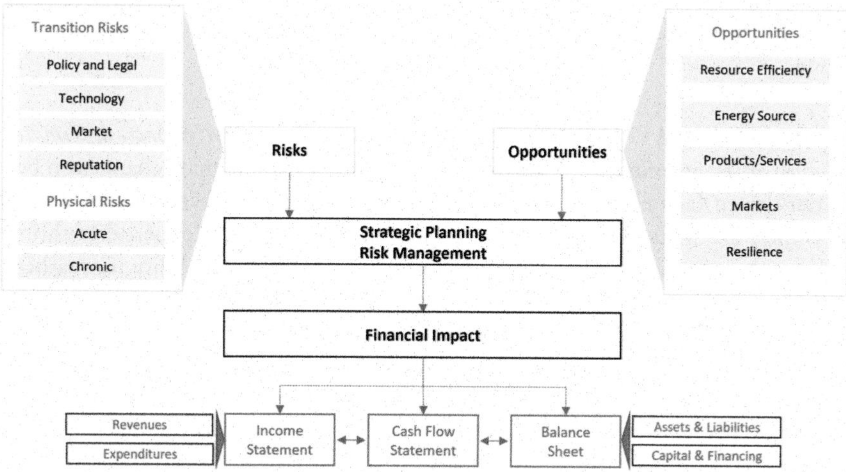

Fig. 3.1 Climate-related risks, opportunities, and financial impact (TCFD, 2017)

Once again, the focus is on the financial impacts of climate-related risks and opportunities, and the concern is the future financial position of an organisation as reflected through its financial statements. There is no direct reference to the impact of the organisation, or the responsibility of the organisation. Indeed, the word 'responsibility' is used 6 times in the entire 74-page report that established the framework, and none refer to the necessity to take responsibility for the space impact of investments, by businesses and the financial sector (TCFD, 2017).

3.2.2 Climate Risks and Climate Responsibilities: Double Materiality

Given our discussion in chapter two and given the theoretical and practical recommendations of the framework, this risk focus is not surprising. Indeed, and of course, the word 'risk' is used 438 times in this report. The word 'time' is used 55 times, and the word space is used 0 times.

Naturally, the mindset of a risk-and-time-focused discipline has affected the logic of the framework recommended by the TCFD. Meanwhile, the entrenchment of responsibility, the integration of space impact into value is yet to be achieved. How appropriate, for a field defined and built around the risk-averse investor, that we are now looking to identify and measure climate risks instead of climate responsibilities. Indeed, the market is now saturated with climate risk narratives (FSB, 2021b, UNEPFI, 2021a, b).

But what are Climate Risks? The 2017 TCFD report defines them as follows:

> CLIMATE-RELATED RISK refers to the potential *negative impacts of climate change on an organization.* Physical risks emanating from climate change can be event-driven (acute) such as increased severity of extreme weather events (e.g., cyclones, droughts, floods, and fires). They can also relate to longer-term shifts (chronic) in precipitation and temperature and increased variability in weather patterns (e.g., sea level rise). Climate-related risks can also be associated with the transition to a lower-carbon global economy, the most common of which relate to policy and legal actions, technology changes, market responses, and reputational considerations (TCFD, 2017, 62).[6]

Interestingly, climate-related financial risks, transition or physical risks, are looked at from the perspective of the stability of the financial system and its players. In other words, risks from climate-related events that could cause systemic shocks or other sector specific shocks. Even Technology is referred to in the risk segment, referring to technology risk, as the world transitions to a low carbon economy. The report describes technology risk as follows:

> TECHNOLOGY RISK Technological improvements or innovations that support the transition to a lower-carbon, energy efficient economic system can have a significant impact on organizations... To the extent that new technology displaces old systems and disrupts some parts of the existing economic system, winners and losers will emerge from this "creative destruction" process. The timing of technology development and deployment, however, is a key uncertainty in assessing technology risk. (TCFD, 2017, 6).

A framework where the climate and the warming of the planet is being treated as a risk to the economy/financial system, and not as a central responsibility of the economy, is addressing only one aspect of the challenge, the more familiar aspect. To attend to the deeper issues, we must consider the impact of the economy on the planet, and make sure businesses and the financial sector are taking responsibility for their impact.

This is not to argue that businesses should not consider the potential risks of floods, fires, and other natural disasters, of course they should. They should also not ignore the risks of transition to a low carbon economy. The point here is that a narrative framework that puts forward voluntary reporting recommendations, that focuses on risk rather than responsibility, more concerned with the impact of climate change on the financial system

---

[6] Emphasis added.

rather than the impact of the financial system on the planet, will soon have to be reinvented.

The two aspects of sustainability, risk and responsibility, have been recently identified and discussed extensively under the broader theme of *double materiality*. The European Commission has addressed the concept of double materiality formally (EU, 2019), and it is neatly introduced and discussed in a report initiated by GRI and authored by Adams et al. (2021). Companies, financial institutions, and standard setters must take both aspects into account. Indeed, this book and the main theoretical and mathematical propositions it puts forward are aimed at addressing this issue in finance theory and practice.

The TCFD framework has been further developed since the publication of the 2017 report and a series of measures and metrics have been proposed and discussed. The main purpose of these metrics is the alignment of portfolios with relevant pathways that would lead to the achievement of our climate targets.

3.2.3 Alignment Through Engagement not Divestment

Recognising the necessary reallocation of capital required, the finance industry is actively engaged in developing tools that help the realignment of capital with the Paris targets.

> Understanding this responsibility, financial institutions are increasingly making public commitments to align their activities with the goals of the Paris Agreement or, more broadly, to reduce their "financed emissions" to net-zero by mid century in a way that is consistent with the achievement of a 1.5 °C target. This is reflected, for example, by the launch of the Glasgow Financial Alliance for Net Zero – GFANZ. These commitments represent a fundamental reshaping of the way that the financial system thinks about allocating capital, which, in turn, is creating a need for new quantitative tools and metrics to govern this process. (TCFD-PAT, 2021, 15)

This paragraph is crucial and very relevant to our discussion. Indeed, this book is entirely about 'reshaping the way that the financial system thinks about allocating capital'. I argue that alignment with tentative pathways through voluntary and flexible frameworks that do not change our existing value equations is not enough to actually change the way the financial system thinks about allocating capital. Indeed, this chapter argues that the metrics of alignment do not go far enough.

When reshaping how capital is allocated, the TCFD report has a very specifically articulated strategic orientation. This is revealed through its position vis-à-vis divestments.

> Specifically, it is critical that the tools and metrics financial institutions use to set climate targets and track progress against them are built to incentivize institutions to engage with counterparties and achieve targets by facilitating their transition, instead of by divesting. It is widely accepted that pursuing divestment will pose substantial problems to the net-zero transition, both on an individual institution level and financial system level, by driving emissive industries out of the regulated capital markets and responsible public ownership, and overinflating demand for already net-zero or post transition counterparties. In other words, only through engagement can financial institutions ensure capital flows toward activities that are aligned with a transition to a 1.5 °C future and is redirected away from those that are not. (TCFD-PAT, 2021, 15)

Even though the report argues for engagement rather than divestment, it goes on to state that building tools and metrics that incentivise 'engagement over divestment' are difficult. The main reasons why this is qualified as difficult are three. First because the emissions of a given entity cannot be assessed alone and must be considered in relative terms to a specific emissions pathway. Second because not every entity or counterparty needs to or can decarbonise at the same rate. Third, because '[p]rojections of the future evolution of counterparty transition performance are necessary so that financial institutions can anticipate when and how specific counterparties are likely to diverge from the needed rate of transition and engage proactively with them to help course-correct' (TCFD-PAT, 2021, 15–16).

While engagement is chosen over divestment, engagement metrics are challenging to build and implement.

3.3 Portfolio Alignment Tools

Amongst the many approaches and measures we observe in this evolving market, the most discussed strategy is the 'Net Zero alignment of companies and investment portfolios'. The TCFD reports by the Portfolio Alignment Team (PAT) identify three 'forward-looking' tools for the purpose. The simplest of them consists in binary target measurement, measuring alignment by the percentage of investments in a portfolio with declared Net Zero/Paris alignment targets. A slightly more complex approach are the benchmark divergence models, which measure alignment at each investment level by comparing asset emissions with a benchmark emission pathway based on forward-looking scenarios. Finally, the Implied Temperature Rise model (ITR), the relatively more complex approach which goes a step further.

> Implied temperature rise (ITR) models: These tools extend benchmark divergence models one step further, translating an assessment of alignment/misalignment with a benchmark into a measure of the consequences of that alignment in the form of a temperature score that describes the most likely global warming outcome if the global economy was to exhibit same level of ambition as the counterparty in question. (TCFD-PAT, 2021, 2)

The report suggests that institutions 'use whichever portfolio alignment tool best suits their institutional context and capabilities'. It also makes several methodological suggestions to all institutions working on developing their benchmarks and metrics. Table 3.3 reveals the key steps and judgements involved in the process.

The report, which is a methodological advisory document to financial institutions with many moving parts and most of the responsibility of design passed on to the institutions, recognises the many limitations of these tools, the most critical of which is data availability. Interestingly, given the voluntary nature of the framework and climate disclosures to date, the issue of data may remain a challenge for the foreseeable future.

Table 3.3 How do portfolio alignment tools work? (TCFD-PAT, 2021)

Methodological step design	Judgement
Step 1 Translating scenario-based carbon budgets into benchmarks	J1: What type of benchmark should be built? J2: How should benchmark scenarios be selected? J3: Should absolute emissions, production capacity, or emissions intensity units be used?
Step 2 Assessing counterparty-level alignment	J4: What scope of emissions should be included? J5: How should emissions baselines be quantified? J6: How should forward-looking emissions be estimated? J7: How should alignment be measured?
Step 3 Assessing portfolio-level alignment	J8: How should alignment be expressed as a metric? J9: How should counterparty-level scores be aggregated?

The next sections will look into specific portfolio alignment tools currently being discussed and used by different entities in the market. While informative, these metrics do not actually integrate impact into value. Furthermore, they also do not offer a decision framework that argues or enforces the rejection of an investment based on a specific score or value.

3.3.1 Carbon Intensity

Carbon intensity is a commonly used measure in the industry. The measure, as used by MSCI (MSCI, 2018) and others, is usually applied to stocks and thus uses sales revenue and emissions in the calculation. The measure calculates the emissions per unit of sales for different investments or assets and portfolios as carbon intensive sales revenues are important to spot and identify when assessing climate risks.

$$Carbon\,Intensity_{Asset} = \frac{Carbon\,Emissions_{Asset}}{Sales_{Asset}} \tag{3.1}$$

While useful for the comparative analysis of assets and the assessment of their carbon dependency, this measure does not actually change valuation models. It does not offer a logical link that affects the valuation of the asset or investment. It is a measure of riskiness more than a measure of value. This is so because a high carbon intensity business or investment will be more directly affected by the ongoing industrial, business, technological and regulatory transition to a Net Zero economy.

The carbon intensity concept is not an alignment tool as such, it is an indicator of exposure, and it has been used as a foundational element in implied temperature rise score calculations by a variety of market providers and institutions.

3.3.2 Implied Temperature Rise

The Implied Temperature Rise (ITR) measure, or temperature score, is a forward-looking metric and aims to assess or identify the global warming contribution of an asset or portfolio. A detailed methodology of translating emission targets into warming contributions is given by CDP-WWF (2020). The CDP methodology, using projected targets, scenarios, and pathways, applies a two-step process to reach end of century temperature outcomes.

In a more recent and detailed methodological paper on temperature scores, the Cambridge Institute for Sustainability Leadership Investment Leaders

Group (CISL, 2021a, 2021b) describes a four-step process. I have chosen to discuss the CISL (2021b) approach here, given the revealing equations at every step. Despite its shortcomings, the CISL ITR model is a commendable contribution in terms of the mathematical transparency it introduces into a debate primarily presented through narrative discussions.

The first step of the process is the measurement of the carbon intensity of the asset/portfolio, then the calculation of its equivalent emissions, then its cumulative emissions, and finally its temperature score. The steps are summarised and described below.

We can calculate the ITR measure using the CISL (2021b) method by first calculating the carbon intensity of the asset:

$$CERI\ Asset = \frac{Scope\ 1\ and\ 2\ Emissions}{Revenue} \tag{3.2}$$

Then measuring the equivalent emissions that calculate the global emissions if the world had the same carbon intensity as the asset:

$$Equivalent\ Global\ Emissions = CERI_{Asset} * Global\ GDP * \theta$$

where Theta is a scaling factor equal to 2.61, calculated through:

$$\begin{aligned}\theta &= \frac{Global\ benchmark\ for\ emissions\ intensity}{Portfolio\ benchmark\ for\ emissions\ intensity} \\ &= \frac{493.18[tCO2/US\$m]}{188.70[tCO2/US\$m]} = 2.61\end{aligned} \tag{3.3}$$

Then calculating the equivalent global cumulative emissions of the asset:

$$Cumulative\ CO_2\ Emissions = \sum_{t=2020}^{2100} Equivalent\ Global\ Emissions_t \tag{3.4}$$

Then finally using the transient climate response to cumulative CO_2 emissions (TCRE) formula to derive the temperature score,

$$Global\ Warming\ Since\ 2020 = \alpha \times Cumulative\ CO_2\ Emissions\ Since\ 2020 + \beta. \tag{3.5}$$

where $\alpha = 5.29 \cdot 10^{-4}$ (°C/GtCO_2) and $\beta = 1.24$ (°C), we find the implied temperature score of the asset.

When looking for the implied temperature rise score of a portfolio, the logic remains the same, but the calculation of CERI differs. It is equal to:

$$CERI_{Portfolio} = \frac{\sum_{i=1}^{n\,assets}\left(\frac{Value\ of\ Investment_i}{EVIC_i} \times Scope\ 1\ and\ 2\ GHG\ Emissions_i\right)}{\sum_{i=1}^{n\,assets}\left(\frac{Value\ of\ Investment_i}{EVIC_i} \times Revenue_i\right)} \tag{3.6}$$

where Value of Investment is the proportion of the portfolio invested in asset i, and EVIC is the Enterprise Value including Cash of the asset i.

$$Value\ of\ Investments_i = Size\ of\ Fund * Weight\ of\ Asset_i \tag{3.7}$$

Enterprise Value including Cash is commonly defined as:

$$EVIC_i = Common\,Shares + Preferred\ Shares + Market\ Value\ of\ Debt + Minority\ Interest \tag{3.8}$$

It is quite clear from the above equations that the temperature score measure does not change the valuation of the asset, it does not change its enterprise value, it simply measures its relative performance in temperature terms using a series of assumptions. In this model, for example, the scaling factor Theta (θ) penalises the climate-reporting assets in the portfolio on behalf of the non-reporting firms in the index used to calculate Theta (CISL, 2021b, 66).

The denominator in the Theta fraction is low, thus leading to a higher Theta, because around 40% of the companies in the index used for the calculation do not yet report emissions. Assuming that they are potentially the ones with much higher intensity, this model is actually ascribing much higher scores to reporting companies simply because of the non-reporting firms, a bias that can send very damaging market signals. Furthermore, it is not clear why this scaling factor has been used, as the model is built on the assumption that it calculates the resulting equivalent emissions if the world economy were to have the same carbon intensity of the asset or the portfolio. Multiplying that figure by 2.61 is not clearly justified.

Moreover, not including scope 3 emissions due to data availability is understandable, but from a modelling perspective, it reinforces the issue discussed above regarding the scaling factor used, Theta. If the scope three emissions of the companies in the index were to be included, alongside the emissions of those who do not report yet, Theta is more likely to be closer to one.

While there are many aspects that can be debated in this model, suffice it to state here that the implied temperature rise measure falls short of transforming the value equations being applied to the assets. Thus, it cannot be used to integrate the impact of an investment into the valuation of the investment. It can hypothetically identify a misaligned asset or portfolio through extrapolations, but it does not describe a realignment, transformation, and transition logic.

3.4 ESG Ratings

ESG scores or ratings have become a mainstream tool used by a variety of investors and asset managers. These ratings are used to assess the compliance and performance of investments in accordance with the new priorities and sustainability objectives. Some asset managers and market intermediaries use them to convince their potential investors that doing good is good for their portfolio value because ESG investments are outperforming other non-ESG investments. Naturally, there are statistically significant results to back these claims (Friede et al., 2015).

Whatever the level of buzz around this acronym, and the number of rating providers, and their profits, ESG ratings fall short of linking to, or integrating with, core finance theory and models. They do not change our existing value equations, and they do not offer or provide a logic through which scores can and should affect value. They also do not measure nor investigate impact. Table 3.4 summarises (verbatim) the scores, grades, and approach used by one of the main ESG score providers, Refinitiv (2021). While the company states that the scores are data-driven, and it does not presume to define what is 'good', through the industry-based materiality weightings of key analytical pillars in each category (E, S, and G), and through the actual data points selected for each, the rater's own interpretation comes through.

The Refinitiv report on their ESG score methodology (2021) provides a detailed view of the firm's approach to calculating the scores. The methodology is discussed through the main text as well as the appendices on pillar scores, pillar weights, categories, materiality matrix, and the type of data points used when assessing the ESG controversy overlay score.

Interestingly, 21 out of 23 data points used for the ESG controversy score start with 'number of controversies published in the media...' (Refinitiv, 2021, 23). Using the media as a source of data to compare 'actions against commitment' reveals yet another important feature of ESG scores. They seem to be scoring what is reported or visible, which is equivalent to scoring

Table 3.4 Refinitiv ESG scores, grades, scoring method (Refinitiv, 2021, 7)

Score range	Grade	Description
0.0 <= score <= 0.083333	D−	'D' score indicates poor relative ESG performance and insufficient degree of transparency in reporting material ESG data publicly
0.083333 < score <= 0.166666	D−	
0.166666 < score <= 0.250000	D+	
0.250000 < score <= 0.333333	C−	'C' score indicates satisfactory relative ESG performance and moderate degree of transparency in reporting material ESG data publicly
0.333333 < score <= 0.416666	C−	
0.416666 < score <= 0.500000	C+	
0.500000 < score <= 0.583333	B−	'B' score indicates good relative ESG performance and above average degree of transparency in reporting material ESG data publicly
0.583333 < score <= 0.666666	B−	
0.666666 < score <= 0.750000	B+	
0.750000 < score <= 0.833333	A−	'A' score indicates excellent relative ESG performance and high degree of transparency in reporting material ESG data publicly
0.833333 < score <= 0.916666	A−	
0.916666 < score <= 1	A+	

> The scores are based on relative performance of ESG factors with the company's sector (for environmental and social) and country of incorporation (for governance). Refinitiv does not presume to define what 'good' looks like; we let the data determine industry-based relative performance within the construct of our criteria and data model. Refinitiv's ESG scoring methodology has a number of key calculation principles set out below
>
> 1. Unique ESG magnitude (materiality) weightings have been included—as the importance of ESG factors differ across industries, we have mapped each metric's materiality for each industry on a scale of 1 to 10
> 2. Transparency stimulation—company disclosure is at the core of our methodology. With applied weighting, not reporting 'immaterial' data points doesn't greatly affect a company's score, whereas not reporting on 'highly material' data points will negatively affect a company's score
> 3. ESG controversies overlay—we verify companies' actions against commitments, to magnify the impact of significant controversies on the overall ESG scoring. The scoring methodology aims to address the market cap bias from which large companies suffer by introducing severity weights, which ensure controversy scores are adjusted based on a company's size
> 4. Industry and country benchmarks at the data point scoring level—to facilitate comparable analysis within peer groups
> 5. Percentile rank scoring methodology—to eliminate hidden layers of calculations. This methodology enables Refinitiv to produce a score between 0 and 100, as well as easy-to-understand letter grades. (Refinitiv, 2021, 3)

'perception management' than actual impact. This is also revealed through some of the non-media-based data points, like having a 'policy on X' or a 'policy on Y' or reporting emissions. In a methodological note on Boolean and numeric data points, the report states:

> For instance, the answer to 'Does the company have a water efficiency policy?' can be 'Yes' (which is equal to 1) if this is indeed the case, or 'No' if the company in question does not have such a policy, or if it reports only partial information (which is equivalent to 'No). (Refinitiv, 2021, 9)

As such, ESG scores are not actually measuring impact, and without actual impact investigations, they could well be called 'ESG public image management' scores. Notice in Table 3.4 how the grades refer to '*relative* ESG performance and ... degree of transparency in reporting material ESG data publicly'.

Discussing the data points used for the ESG scores, the report states:

> Refinitiv captures and calculates over 500 company-level ESG measures, of which a subset of 186 (details in the ESG glossary, available on request) of the most comparable and material per industry, power the overall company assessment and scoring process. (Refinitiv, 2021, 6)

While the 23 data points used for the ESG controversies score are listed in the Appendix G of the report, discussed above, the 186 data points through which the ESG scores are calculated are available upon request. I do not recall ever reading a methodological paper in finance where the actual data points used were available upon request. We do not need the actual data, but the data points used are critical to assess the scores, and yet they are not included in the ESG scoring methodology paper (Refinitiv, 2021).[7] We need all 186 data points listed to understand what the methodology actually is, and what it actually measures.[8]

Similarly, in a recent methodology paper on their ESG scores, S&P Global reveals (S&P Global, 2022b) a questionnaire-based scoring system. The 'Corporate Sustainability Assessment' is an annual assessment with 61 industry-specific questionnaires. When companies do not provide answers, S&P's team of experts fill them up based on their own research. The report describes the process as follows:

> ESG Scores are measured on a scale of 0–100, where 100 represents the maximum score. Points are awarded at the question-level (on average 130 per company) based on our assessment of underlying data points (up to 1,000

[7] I would like to state clearly that Refinitiv is one example, and other ESG rating providers have similar practices, and their methodologies face similar issues (MSCI, 2020).

[8] Note that I have not requested the glossary and the data points, and the argument here is not about their eventual possible availability. The point of concern here is the fact that the main methodology paper on ESG scores by one of the main ESG score providers does not openly disclose the data points.

> per company) according to pre-defined scoring frameworks that assess their availability, quality, relevance, and performance on ESG topics. (S&P Global, 2022b, 3)

Describing how a specific question affects the final ESG score of a company, the report describes a hypothetical example and the process in Appendix I as follows:

> In 2021, the environmentally oriented bank, Blue Sycamore Bank, participated in the S&P Global Corporate Sustainability Assessment (CSA). Due to their comprehensive environmental impact reporting practices, they received 100(/100) points on the 'Environmental Reporting – Coverage' *question.* Within the 'Environmental Reporting' *criteria* for the 2021 Banks CSA, the 'Environmental Reporting – Coverage' question has a weight of *50%.* Therefore, the 'Environmental Reporting – Coverage' 100 question points contributed 50 points (100 * 0.5) to the 'Environmental Reporting' criteria. Within the 'Environmental' *dimension* for the 2021 Banks CSA, the 'Environmental Reporting' criteria has a weight of *23%.* Therefore, the 'Environmental Reporting – Coverage' 100 question points contributed 11.5 points ((100 * 0.5) * (0.23)) to the 'Environmental' dimension. Within the S&P Global *ESG Score* for the 2021 Banks CSA, the 'Environmental' dimension has a weight of *13%.* Therefore, the 'Environmental Reporting – Coverage' 100 question points contributed 1.5 points ((100 * 0.5) * (0.23) * (0.13)) to the S&P Global ESG Score. (S&P Global, 2022b, 13)[9]

We can see from the above that given the score and weights at each level of aggregation (Question, Criteria, Dimension, ESG Score), and the number of initial questions, the final score is a subjective summary of many different small and large, significant and not so significant data points, many of which describe what is reported and visible, without necessarily investigating or measuring actual impacts.

The S&P ESG scoring methodology paper, while different from the Refinitiv report as it does reveal the criteria and questions used for different dimension scores in Appendix III (S&P Global, 2022b, 19), it only reveals the broad category criterion and title of the questions, not the actual questions, which means that we cannot gauge the possible answers (data points) or the exact meaning of the questions.[10]

---

[9] Emphases added to denote the layers of aggregation and weights being applied.

[10] For example, under the Environmental Dimension, we have a 'Packaging' criterion described with a 'Packaging Commitment' question (S&P Global, 2022b, 19).

Studies have also shown that there is a 'disagreement' between popular ESG ratings. In other words, the ratings given by different providers do not always correlate, and scores available in the market for the same stocks differ (Berg et al., 2022; Christensen et al., 2021). Gibson et al. (2020) find that stock returns are actually positively related to ESG rating disagreement, and that this is mainly based on disagreement about the environmental aspect of scores. Based on the evidenced divergence of ESG ratings, a number of studies have also looked at their relationship to returns (Avramov et al., 2021).

Berg et al. (2022) reveal ESG ratings divergence by looking at the ratings of the top six providers, i.e., KLD, Sustainalytics, Moody's ESG, S&P Global, Refinitiv, and MSCI. They suggest that the measurement is responsible for 56% of the divergence, suggesting what they call a 'rater effect'. They also raise the necessity to pay greater attention 'to how the data underlying ESG ratings is generated'. Moreover, based on a series of industry and academic papers, there are a number of biases in ESG scores, namely, market capitalisation, country of origin, and sector (Ratsimiveh & Haalebos, 2021; Ratsimiveh et al., 2020). ESG ratings are used as descriptive scores, often linked to risk, even though their methodology is not always transparent nor always comparable (Inderst & Stewart, 2018; Sherwood & Pollard, 2019). Furthermore, there are a number of challenges related to the E score of ESG ratings (Boffo & Patalano, 2020; Boffo et al., 2020).

In their OECD report, Boffo et al. (2020) state:

> Also, while the E score includes a number of distinct environmental metrics, the analysis found a positive correlation between some ESG raters' high E scores of corporate issuers and high levels of carbon emissions and waste. (Boffo et al., 2020, 7)

The main reason behind the many issues we face with ESG ratings stem from the fact that ESG ratings are opinion points rather than data points. If ESG ratings were data points, they would be provided to the market exactly and in the same frequency that other types of data are fed into the market. They are similar to credit ratings in some ways, with a key difference being the lack of any established theoretical framework for their assessment. Debts and credit have a formally recognised valuation theory that the market understands and recognises.

ESG scores, besides being infrequent and incomparable, they also do not provide the market with a logic and associated equations through which an investor/trader can assess a specific new data point that affects the ESG score/performance of an asset. The packaged interpretations are delivered to the market, and however frequent they become, they fall short of providing a mechanism and logic through which incremental information can be assessed by a trader or investor, in between the publication of scores and their updates.

The reason why ESG ratings have become popular is because they help subcontract the most tedious and most complex step of impact assessment and provide 'scores' that can be used to keep everything working as they always have. Whatever the specific methodology used by different ESG ratings, the fact that they are interpretations rather than data points raises a few fundamental questions. If the market is subcontracting the hardest task of impact measurement and interpretation to ESG rating providers, then, in truth, once again, it is externalising the treatment of impact from its core processes. This implies that ESG scores, however useful, are indirectly contributing to the marginalisation of sustainability from core finance theory and application.

Interestingly, ESG scores do not *actually* measure impact, and seem to be perception-based ratings that are offered as packaged interpretations where the methodology and actual rationale of a specific judgement change is not immediately evident to the market. As the following chapters will discuss and reveal, the models and solutions proposed in this book also require some form of judgement and interpretation by investors. The point being made here is that with ESG ratings there is no clear framework or a set of clearly defined equations through which investors can make up their mind and adjust their judgements as new information becomes available. The packaged nature of interpretations and the lack of clear interpretive frameworks that can allow the analysis of new incremental information is the issue.

Although popular with banks, central banks, and asset managers, ESG ratings or scores, even when useful, do not integrate the impact of cash flows and assets into the valuation models of those assets or cash flows. Furthermore, they are not readily compatible with existing discounted cash flow models or other asset pricing models without theoretical and mathematical adjustments that are not part of the offering of ESG ratings.

Moreover, for a market that has been digitised long before sustainable finance and ESG became so popular, algorithmic trading cannot be based on infrequent and non-comparable scores. As such, the currently intensifying digital transformation will require that we eventually replace ESG ratings with ESG algorithms. Naturally, availability of data will be key to achieve this transformation, but before the data points, we will need the equations that will make decision algorithms possible. More on this in chapter seven.

In the next section, I explore the actual strategies through which asset managers integrate ESG factors in their decision-making process.

3.5 ESG Integration in Practice

This section is primarily based on a relatively recent research report published by the Principles of Responsible Investment (PRI, 2016) in collaboration with the United Nations Environment Program (UNEP) Finance Initiative (UNEPFI) and United Nations Global Compact (UNGC). The report explores ESG integration in practice through actual examples from a variety of asset managers, all reported directly by the asset managers themselves.

The report defines ESG integration as:

> The PRI defines ESG integration as "the systematic and explicit inclusion of material ESG factors into investment analysis and investment decisions". It is one of three ways to incorporate responsible investment into investment decisions, alongside thematic investing and screening. All three ESG incorporation practices can be applied concurrently (PRI, 2016, 12).

ESG factor integration is discussed in the context of four different types of investment strategies: (1) Fundamental, (2) Quantitative, (3) Smart Beta, and (4) Passive. The integration model described goes through 4 stages: (a) qualitative analysis, (b) quantitative analysis, (c) investment decision, and (d) active ownership assessment.

The ESG integration strategies and examples often use clip-art tables and drawings to demonstrate the many aspects of E and S and G and how they should be integrated, but in the end, the value equations and the value paradigm discussed in chapter two remains intact. Instead, ESG integration is tantamount to adjusting conventional valuation variables in conventional valuation models. As per the PRI report (2016), integrating ESG factors in fundamental investment strategies involves the following:

> Forecasted company financials drive valuation models such as the discounted cash flow (DCF) model, which in turn calculates the estimated value (or fair value) of a company and hence can affect investment decisions. Investors can adjust forecasted financials such as revenue, operating cost, asset book value and capital expenditure for the expected impact of ESG factors (PRI, 2016, 13).

This, briefly and yet clearly, summarises the core issue. The financial valuation models themselves are not affected. In other words, the use of the discounted cash flow models remains central in deriving the fair value of a stock or cash flows, and the relevance of ESG factors is determined through the impact they would have on the forecasted variables used in the DCF models. This approach, in other words, uses the same valuation logic described in the previous chapter, but adjusts the forecasted financials to potential impacts from ESG factors.

Looking at Quantitative investment strategies, the PRI (2016) report states:

> The quant managers that perform ESG integration have constructed models that integrate ESG factors alongside other factors, such as value, size, momentum, growth, and volatility. ESG data and/or ratings are included in their investment process and could result in the weights of securities being adjusted upwards or downwards, including to zero. There are two main approaches to integrating ESG factors into quantitative models. They involve adjusting the weights of: securities ranked poorly on ESG to zero, based on research that links ESG factors to investment risk and/ or risk-adjusted returns; each security in the investment universe, according to the statistical relationship between an ESG dataset and other factors (PRI, 2016, 36).

The integration of ESG factors into portfolio weighing decisions, i.e., the decisions to increase, decrease, or remove an asset from the portfolio, is done through proprietary models that are not discussed or shared. Interestingly, worth noting that ESG factors are being entered into these models testing price and return correlations 'alongside other variables like value, size, momentum, growth, and volatility'.

While ESG factors or ratings are variables alongside volatility and size, ultimately affecting investment weights in the portfolio, there is no logic revealed. Would a high positive return stock be removed if its ESG factors reveal non-compliance? As the report states, the weights are adjusted 'based on research that links ESG factors to investment risk and/or risk-adjusted returns'.

We are back to where we started. Given what we have learned in chapter two, and the discussion about ESG ratings in the previous section, the integration approach that links back to risk-adjusted returns does not involve any change in our value paradigm or equations.

On Smart Beta strategies, the PRI (2016) report states:

> In smart beta strategies, ESG factors and scores can be used as a weight in portfolio construction to create excess risk-adjusted returns, reduce downside risk and/or enhance portfolios' ESG risk profile (PRI, 2016, 43).

It is quite revealing that the first possible purpose mentioned by the report is 'to create excess risk-adjusted returns'.

Discussing ESG integration in passive strategies, the report states:

> One approach is to reduce the ESG risk profile or exposure to a particular ESG factor by tracking an index that adjusts the weights of constituents of a parent index accordingly (PRI, 2016, 50).

If ESG integration is about 'adjustments of weights' in portfolios, which based on the many examples and conceptual strategies discussed, it is, then it does not actually imply any change in valuation. Furthermore, adjustment of weights from 50 to 10% does not involve any change in emissions in the company being 'weighted down'. Trading down the weights of an asset in a portfolio involves selling the stock or asset, which also implies that someone else is buying it. This means that ESG integration has less impact on the valuation of stocks and assets, and more on the holding demographics of stocks. Even when the weight of an asset is reduced to zero, this analysis applies. Unless, of course, the stocks being weighted down or removed from portfolios, i.e., weights adjusted to zero, are also experiencing a decline in prices due to adjusted valuations.

Coupled with a TCFD framework that advocates engagement rather than divestment, we are left to wonder how effective a transition based on these 'adjustment of weights' can be.

As Tariq Fancy, the former Chief Investment Officer for Sustainable Investing at BlackRock writes:

> In my role at BlackRock, I was helping to popularize an idea that the answer to a sustainable future runs through ESG and sustainability and green products, or in other words, that the answer to the market's failure to serve the long-term public interest is, of course, more market (Fancy, 2021).

While Fancy (2021) addresses a number of relevant issues to the way 'ESG' is being integrated into financial products and their marketing, the point here is that it is not enough to use ESG factors, through ratings or other, to adjust variables in the same models we have been using on our way to the current crisis. As I have argued previously and will continue to argue in this book, the issue is not a 'market failure' the way the term has been understood in economics and described by Fancy (2021). Our challenge is more serious and runs deeper.

In order to truly entrench sustainability into finance theory and practice, we need a transformed financial value paradigm, adapted principles and equations, and, as chapter eight will argue, a commensurately reformed monetary architecture. Chapters four, five, six, seven, and eight discuss this transformation in detail.

3.6 Responsible Banking, Nature-Related Financial Disclosures, and Impact Investing

The accounting standards and frameworks, the portfolio alignment tools, and the ESG ratings discussed to this point do not exhaust the many approaches and initiatives within the broader sustainable finance landscape. While it is outside of the scope of this chapter and book to review all and every initiative in the field, a few deserve to be discussed here. The first one is impact investing, and the other two, albeit further away in terms of any structured solutions, are the Principles of Responsible Banking (PRB) and the Taskforce on Nature-related Financial Disclosures (TNFD).

3.6.1 Impact Investing

In the currently active and growing field of impact investing (Agrawal & Hockerts, 2018; Nusseibeh, 2017; Bertl, 2016; Bugg-Levine & Emerson, 2011; Jackson, 2013; Reeder et al., 2015), the challenge of a clear and replicable methodology of measurement is clearly discussed and identified. Bugg-Levine and Emerson (2011) have put forward the concept of Blended Value combining the economic, social, and environmental components of value (2011, 10). But, as they describe, '[d]espite this new attention, the fundamental challenge remains unresolved: How do we develop a measurement system that offers an integrated understanding of blended value creation that matches the interest of the impact investor?' (Bugg-Levine & Emerson, 2011, 167).

The measurement of impact has attracted significant attention, and since the five capitals framework proposed by the Forum for the Future (FFTF, 2012) and the six capitals framework proposed by the IIRC (IIRC, 2013, 2021), and the three capitals inclusive wealth concept of UNEP (2018), a variety of approaches have been devised to structure the measurement of impact according to these capitals (Fig. 3.2).

Over the last decade or so, many governmental and non-governmental organisations, industry associations, and research institutions have made their own contributions to the field of impact investing. The Impact Management Project (IMP, 2016) focuses on measuring the 'five dimensions of impact:

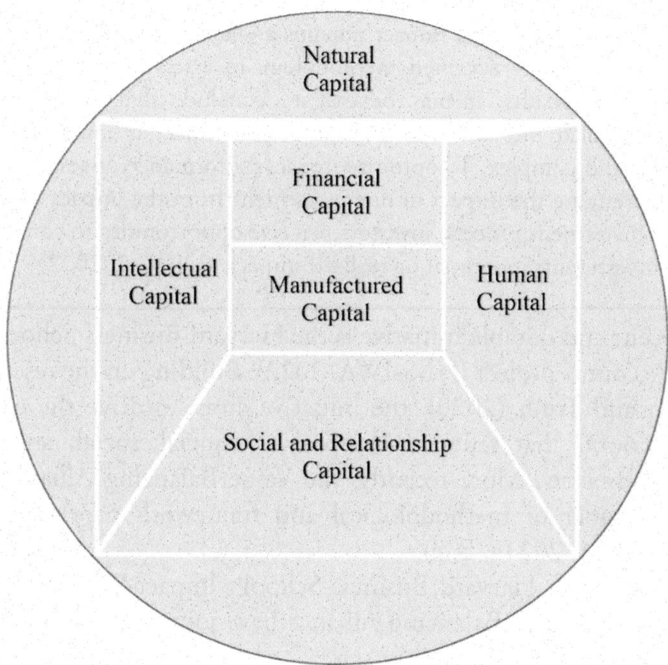

Fig. 3.2 The Six Capitals (Adapted by author from IIRC, 2013)

what, who, how much, contribution, and risk'. The Impact Institute (II, 2019) has published its own framework for 'impact statements'.

A recent report by the Global Impact Investing Network (GIIN, 2016) discusses the different ways impact investors derive value from impact measurement and management. While the report does not include any equations for measurement, it does give procedural advice as to how investors can choose targets and achieve the impact they desire given risks and stakeholders involved. The GIIN refers to the Social Impact Investment Taskforce report titled 'Measuring Impact' for the definition of the impact measurement and management process. According to the SIIT report (2014), the impact measurement and management process involves four stages: (1) plan, (2) do, (3) assess, and (4) review (SIIT, 2014).

Interestingly, while there are many framework and stakeholder flowcharts across the reports, equations for actual impact measurement in monetary terms are still missing, and none offer any procedural recommendations through which impact can be integrated into our existing value equations in core finance theory and practice.

The GIIN in a follow up report (2020) recognises this methodological and measurement challenge presented by impact, and states:

> Yet gauging investment-level impact remains a challenge: until now, there has been no tested, widely accepted methodology to assess and, most critically, compare impact results. It may be easy to conclude that impact can and should be distilled into a monetary figure, yet impact is inherently multi-dimensional and complex. To optimize impact performance, investors must be able to differentiate the impact of one investment from the impact of another. Across the investment process, investors perceive opportunities to compare and optimize investments' potential or realized impact. (GIIN, 2020, 2)

Another recent and notable initiative is the Harvard Business School Impact-Weighted Accounts project (HBS-IWA, 2022). Building on the research done by Serafeim and Trinh (2020), the initiative aims 'to drive the creation of financial accounts that reflect a company's financial, social, and environmental performance'. More recently, the Value Balancing Alliance has put forward a number of methodological and framework papers on 'Impact Statements' (VBA, 2021a, b, c).

Very recently, the Harvard Business School's Impact-Weighted Accounts initiative and the Value Balancing Alliance have joined forces. In their joint statement (HBS-VBA, 2022), they state:

> Today, we recognize and welcome the international harmonization and standardization of frameworks for sustainability reporting and disclosures. However, the landscape of methodologies to assess corporate sustainability performance is still fragmented. To ensure robust and comparable information, efforts to standardize the definition, measurement, and valuation of positive and negative impacts from business on society and enterprise value need to be intensified.
>
> Several methodological approaches are currently being explored. Most fail to handle the measurement and integration of economic, social, and environmental aspects at the same time in a consistent way. Harvard Business School Impact Weighted Accounts (HBS IWA) and the Value Balancing Alliance (VBA) are convinced that monetary valuation of impacts on society and enterprise value is the most promising way to further develop respective sustainability accounting systems in a meaningful, fast, and business compatible manner. (HBS-VBA, 2022)

While the necessity to measure impact and integrate it into value and investment decisions is established, an effective framework of measurement and integration is still work in progress.

3.6.2 Principles of Responsible Banking

Developed by thirty founding banks in collaboration with the United Nations Environment Program Finance Initiative (UNEPFI) through public consultations involving many other stakeholders, the Principles of Responsible Banking were officially launched in 2019. Since then, the number of signatories has risen to 270, representing more than 45% of global banking assets. Describing the principles, the UNEPFI website states:

> *The Principles for Responsible Banking* are a unique framework for ensuring that signatory banks' strategy and practice align with the vision society has set out for its future in the Sustainable Development Goals and the Paris Climate Agreement. (UNEPFI, 2021a, 2021b)

I would like to draw the reader's attention to the key word in the above definition, i.e., 'align'. We do not seem to have any challenges verbalising and setting the high-level targets and objectives that would bring about our ideal developments. The challenge comes down to the actual will and responsibility to do so with the necessary urgency and effectiveness. This framework is a voluntary framework, and it faces the same key issue we have identified and discussed previously in this chapter. It does not transform the many value equations and principles used by banks when lending and valuing debt instruments. As such, while a positive development, they cannot introduce the type of change we need to meet the challenges we face.

The principles (Table 3.5) are six in number and are 'designed to bring purpose, vision, and ambition to sustainable finance', and the signatory banks 'commit to embedding these six principles across all business areas, at the strategic, portfolio and transactional levels' (UNEPFI, 2021a, 2021b).

While I have only admiration for such an effort, and any bank that commits to abide by such principles, I would like to remind the reader that the 2008 Financial Crisis happened just fourteen years ago and it has not been a decade since the days when many of these banks had to settle multi-billion lawsuits for their misconduct before and during the crisis. Moreover, as discused in the introduction of this book, 2021 was a record year of fines for UK's Financial Conduct Authority.

In December 2021, UK's Financial Conduct Authority (FCA) and Prudential Regulation Authority (PRA) fined three major banks for different failings. Standard Chartered was fined £46.5 million by the PRA (Bank of England, 2021) for failing to be open and cooperative and for failings in its regulatory reporting governance and controls. HSBC was fined £63.9 million for deficient transaction monitoring controls (FCA, 2021b), NATWEST was fined

Table 3.5 Principles of Responsible Banking (UNEPFI, 2021a, 2021b)

1-Alignment	2-Impact & target setting	3-Clients & customers
We will align our business strategy to be consistent with and contribute to individuals' needs and society's goals, as expressed in the Sustainable Development Goals, the Paris Climate Agreement and relevant national and regional frameworks	We will continuously increase our positive impacts while reducing the negative impacts on, and managing the risks to, people and environment resulting from our activities, products and services. To this end, we will set and publish targets where we can have the most significant impacts	We will work responsibly with our clients and our customers to encourage sustainable practices and enable economic activities that create shared prosperity for current and future generations
4-Stakeholders	**5-Governance & culture**	**6-Transparency & accountability**
We will proactively and responsibly consult, engage and partner with relevant stakeholders to achieve society's goals	We will implement our commitment to these Principles through effective governance and a culture of responsible banking	We will periodically review our individual and collective implementation of these Principles and be transparent about and accountable for our positive and negative impacts and our contribution to society's goals

£264.8 million for anti-money laundering failures (FCA, 2021a). Earlier in the year, in October 2021 the FCA fined Credit Suisse (FCA, 2021c) over £147 million for 'serious financial crime due diligence failings related to loans worth over $1.3 billion'. In July 2021, the FCA fined Lloyds Bank General Insurance Limited, £90 million 'for failing to ensure that language contained within millions of home insurance renewals communications was clear, fair and not misleading' (FCA, 2021d). As of today, all these banks except HSBC are listed as signatories of the Principles of Responsible Banking (UNEPFI, 2022).

Indeed, in the many legal documents and regulatory statutes and standards that govern the banking sector, there are many legally binding clauses that clearly set professional standards and prohibit the type of conduct that led to the 2008 financial crisis and the many millions of fines described above. These failings have breached actual laws not voluntary principles or frameworks.

While we should remain hopeful that these principles will be adopted and implemented, we must understand them within the context of the previous discussion and the theoretical context and the risk/time value framework within which the industry operates. Indeed, our entire monetary architecture, where banks have a central role, is a contributing factor to the current crisis. The risk/time focus of the value of money is intricately linked to the debt-based methodology of money creation, also intricately linked to banks. I discuss this issue in more detail in chapter eight.

3.6.3 Impact on Nature and the Taskforce on Nature-Related Financial Disclosures

The TNFD or the Taskforce on Nature-related Financial Disclosures was announced in July 2020 created and brought to life by a wide collection of governmental institutions, international institutions, banks, asset managers, consulting firms, think tanks, and others. The purpose, as it is mentioned on the home page of the initiative is to deliver 'a risk management and disclosure framework for organisations to report and act on nature-related risks' (TNFD, 2021a).

Much like the TCFD framework, it seems the TNFD is also concerned with the impact of nature on our financial system, rather than the responsibility of our financial system and its impact on nature. I discussed this lopsided approach in previous sections. Indeed, this is not a hidden agenda, it is stated by the TNFD on its website and first publications (TNFD, 2021a, 2021b, 2021c).

> Nature Loss Presents Financial Risks to Organisations: More than half of the world's economic output – US$44tn of economic value generation – is moderately or highly dependent on nature. Nature loss therefore represents significant risk to corporate and financial stability. Financial institutions and companies need better information to incorporate nature-related risks and opportunities into their strategic planning, risk management and asset allocation decisions. (TNFD, 2021a)

The TNFD is still working towards the development of its framework, after a number of expected milestones in 2022 and a series of consultations, the final release is expected in 2023. Thus, I believe any further discussion must be contingent on the release of the final framework.[11] Given the published statements, however, the risk-focused approach is evident.

A parallel effort in exploring the impact on nature and biodiversity is the report published by the Cambridge Institute of Sustainability Leadership, the Natural Capital Impact Group (CISL, 2020). The report discusses the importance of nature and biodiversity impact and risk assessment. It proposes a Biodiversity Impact Metric (BIM) which is aimed to help and support businesses in measuring the impact of their sourcing agricultural materials in global supply chains, in identifying high-risk locations, in informing the development strategies, and aligning with global goals and science-based targets like the ones by IPBES (IPBES, 2019).

The proposed metric is given below:

$$\textit{Biodiversity Impact Metric} = \textit{BIM}$$

$$\textit{BIM} = \textit{Land Area} \times \textit{Proportion of Biodiveristy} \times \textit{Biodiversity Importance} \quad (3.9)$$

Describing the metric, the report states:

> The outputs from the metric are most easily interpretable in relative terms, for example, by examining whether the sourcing of a commodity is having a higher or lower impact on biodiversity per tonne sourced compared to other sourcing locations or the global average. By examining the total weighted hectares, a company can assess where their greatest exposure to sourcing risk might lie. (CISL, 2020, 11)

While a very timely and appropriate conceptualisation for the purposes of the report defined above, the metric is a comparative metric, and once again, does not go as far as integrating the biodiversity impact of investments into the value equations of investments.

Following the recent publication of the CBD (2021) global framework for managing nature through 2030, a very recent note by the UK's Parliamentary Office of Science and Technology (UK POST, 2022) identifies and addresses the financial risks of nature loss and the alternative approaches to their assessment and mitigation. The note recognises the challenges of measuring and predicting biodiversity loss for the finance sector, and argues for standardised biodiversity disclosures, policy and regulation.

---

[11] On the 15th of March 2022, TNFD released its beta framework for consultation. The final recommendations are due to be released in Q3 2023 (TNFD, 2022).

Before drawing the main conclusions of this chapter, I believe it is necessary to discuss the market shaping sustainable finance and the recent focus on ESG.

3.7 The Market Shaping Sustainable Finance and ESG Reductionism

Across all the initiatives discussed, the main strategy used by the institutions involved in the development of the new standards, frameworks, and metrics is to 'reveal the standards' by combining proprietary analysis and publications with public consultations and debate, referencing and relying on the practices of large industry players as building blocks.

While inclusive and balanced, this approach to the standards and frameworks of sustainable finance poses a challenge, it can potentially water down the transition before we begin. This is so for two main reasons. The first one, as discussed earlier, is the fact that until recently finance expertise has been built on the axiomatic belief in the maximisation of risk-adjusted returns over time, without much thought about impact. The second one is the fact that a quick survey of the market reveals a diverse topography of attitudes vis-à-vis sustainability. Without going too deep into each and without giving any examples, I discuss the most prominent ones I have encountered.

Greenwashing: Given changes in public opinion, greenwashing is one such attitude. Adjusting rhetoric and marketing to this new reality irrespective of the level of actual measurement, compliance, or concern for sustainability within organisations and/or portfolios.

Safe Play: Given changing trends, but operating within the older value framework, safe play is another dominant attitude. Still prioritising the risk-averse investor, this approach aims to convince potential investors and market participants, often with statistically significant results, that sustainable investments are outperforming other non-sustainable investments in the market—so they are within the investors' risk/return optimisation targets. Thus, sustainability is presented as a profitable option, without entrenching responsibility into the investment function.

Denial: Climate change denial cuts across different societal groups, and this attitude is also present in the finance industry. This attitude sees no need to adjust strategies, models, and even rhetoric. This attitude, right or wrong, is quite resistant to change. Given that climate-related disclosures are still not mandatory, the debate is ongoing and there seems to be no immediate pressure to change.

Abstention: Observing the lack of clear principles and methods, the diversity of interpretations, the variety of strategies, and the fact that climate disclosures are not yet mandatory, abstention is another common attitude. A wait-and-see strategy that does not take any clear or active position.

Leadership: Last but not least, we observe a real sense of responsibility and leadership advocating for change. In some instances, this is evident even before the current rise to popularity of sustainable finance. Engaging in active stewardship and leading by example, this approach has been driving the growth in sustainable finance, in theory and practice.

While many distinguished institutions and organisations have consistently advocated for sustainability, even before its current popularity, this topography of attitudes must be considered when our standards are being revealed by and through the market.

With a diverse set of players and levels of commitment and understanding in the market, the need to catalyse the development of robust sustainability standards and tools has never been stronger. As the standards and tools of sustainable finance are revealed, debated, and shaped, it is necessary to make sure we do not dilute their rigour and effectiveness before the transition has had time to take hold.

Given increasing momentum, probably due to convenience, we can see that the market seems to be converging towards ESG factor integration, often through ratings and scores, built in house or purchased in the market. While E, S, and G are important and relevant aspects of sustainability, they do not really exhaust the many facets of the challenges we face.

I mention and discuss four of them here—given their relevance and actuality. These aspects are further developed and discussed in later chapters.

3.7.1 ESG, Responsible, Climate, and Sustainable Finance

The first most important issue in the development of standards in sustainable finance is the necessity to address the seemingly benign issue of naming the field. Given the initial phase of growth and diversity in the approaches applied, we are faced with many names and titles being used to describe the content of the practice or subtle variations of it. Although this describes a positive reality, diverse interests and enthusiasm from a variety of groups and approaches, from a long-term perspective of building a coherent body of knowledge that can actually save the planet and our future in it, it is not necessarily so.

In a recent report PRI (2020) states:

> There are many terms—such as sustainable investing, ethical investing, and impact investing—associated with the plethora of investment approaches that consider ESG issues. Most lack formal definitions, and they are often used interchangeably. A key to understanding how responsible investment is broader than these concepts is that where many make moral or ethical goals a primary purpose, responsible investment can and should be pursued by the investor whose sole focus is financial performance, as well as those looking to build a bridge between financial risk/opportunities and outcomes in the real world. (PRI, 2020, 4)

PRI (2020) advocates for the term Responsible Investment. Indeed, responsibility is a central theme of this book, and the many concepts presented in the following chapters. However, it is my personal conviction that the right strategic choice is to make finance sustainable and responsible. It is not about a subtitle of the field, but about integrating responsibility into the very core of finance.

As we develop standards for sustainable finance to address the climate crisis, we will come to the realisation that what is really needed and currently missing is a framework of value built around responsibility. Otherwise, soon after addressing Environmental, Social, and Governance factors, we will need to address a host of others that we have ignored because of the interpretive focus we have ascribed to the field.

Responsibility must complement the existing value framework of finance theory and practice. Responsibility is key but it is not the purpose. The end goal is to provide us with a sustainable trajectory where we can survive and thrive on this planet and beyond, and that can be achieved when finance becomes inherently responsible and sustainable. The introduction of the space value of money principle aims to contribute to such a change, entrenching responsibility into our core finance theory and practice. I discuss and introduce the principle in chapter four.

3.7.2 Space Layers and Sustainable Finance

During the last 60 plus years, with the expansion of outer space exploration, our experiments, successes, and failures have left a trail of debris orbiting the planet.[12] It seems, our lack of responsibility has also caused extensive pollution of a different kind beyond our atmosphere.[13]

---

[12] For a recent report on space debris, or space junk please see ESA (2021).

[13] *Image Source* https://orbitaldebris.jsc.nasa.gov/photo-gallery/_images/highresolution/geo-2019-4096.jpg.

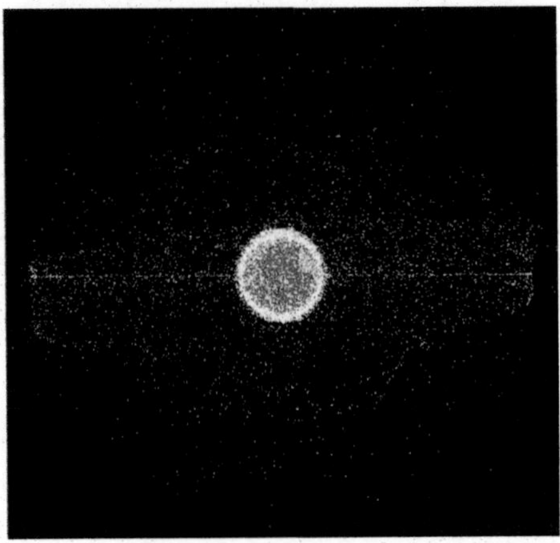

Image Credit NASA

'There are approximately 23,000 pieces of debris larger than a softball orbiting the Earth. They travel at speeds up to 17,500 mph, fast enough for a relatively small piece of orbital debris to damage a satellite or a spacecraft. There are half a million pieces of debris the size of a marble or larger (up to 0.4 inches, or 1 cm) or larger, and approximately 100 million pieces of debris about 0.04 inches (or one millimeter) and larger. There is even smaller micrometer-sized (0.000039 of an inch in diameter) debris' (NASA, 2021).

A robust and effective framework of sustainability must adopt a wider perspective. Sustainable finance must transcend the narrow or just terrestrial interpretations of our context and develop principles and standards that can hold everyone to account, including aerospace companies flying to the moon and Mars. This is more relevant than ever, given the growth in space tourism and the aggressive drive towards further outer space colonisation.

Similarly, when we take responsibility to preserve our terrestrial environment around us by reducing emissions, that sense of responsibility and the standards that enforce it must extend beyond emissions as such. From the microscopic world of atoms and subatomic particles to our cities and rivers, and the moon, our sense of responsibility must stretch across all. Naturally, the pollutants we are looking for and their impact may differ, toxicity may mean different things in different contexts, but from a finance perspective, responsibility to safeguard the environment in its broadest sense remains. Sustainable finance, therefore, through its standards and models, must be able to address investments across all these aspects.

Thus, we must conceptualise and account for our impact on space and its many layers. Indeed, when we think of the dimension of space, we must consider our physical context, stretching from the quantum world of atoms to the gravitational world of cosmology (Smolin, 2006).

I discuss and explore these and related issues in chapters four, five, and six.

3.7.3 Technology and Sustainable Finance: AI, Algorithms, and Digital Data

The role of digitisation in securing the continued and successful growth of sustainable finance is often misunderstood and exploited as part of the buzz around both. Indeed, given the digitisation of entire industries and their value chains, of markets and trading, the key to maintaining the rise of sustainable finance is to ensure its principles, models, and application are integral to the digital transformation we are going through.

As mentioned briefly in the preface of this book, the digitisation of a process does not automatically mean an improvement of that process. An inefficient process can remain so after digitisation. An unfair process can remain so after digitisation. The theoretical and structural design of our valuation models do not change or improve automatically through or after digitisation. They change when we reinterpret them in their fundamental assumptions, and rebuild them upon a transformed value framework leading us to new equations. Moreover, when a theory or logic is fuzzy or unclear, digitising it does not make it clearer and more rigorous.

Given the current levels of theoretical clarity, the digitisation of sustainability in finance is still at the stage of alchemy. It is thus imperative to address the multifaceted evolving nature of impact measurement in finance theory and practice first. I address this topic in more detail in chapter seven.

3.7.4 Money Mechanics and Sustainable Finance: Debt vs Crypto

Another aspect that seems to be overlooked through an ESG focus is money (M). The role of debt-based money in creating and instigating the growth craze that has driven and still drives the ever-expanding exploitation of our ecology is an important aspect to explore. Public and private debts are the cornerstone of money supply growth and expansion, and the instruments used by central banks and banks, and their impact, are crucial for any effective change. Indeed, alongside private and public investors, money creators should also have shared space-based responsibilities, and our principles and standards must be able to address them.

On the other side of the spectrum, cryptocurrencies, theoretically challenging debt-based money and growing in popularity, have a very high carbon footprint, and while they resolve the debt aspect of fiat money, they exponentially increase the carbon problem. Whatever their status, whether we think they have intrinsic value or not, whether we consider them assets or currencies, we face the necessity to build frameworks, standards, and tools that address these issues.

Once again, the digitisation of a process does not imply the improvement of that process. Digital processes can be just as inefficient and unproductive as non-digital processes, and we must address these issues in the context of money and money creation if our efforts to integrate sustainability are going to lead to tangible change. I explore these issues in more depth in chapter eight.

3.8 Conclusion

This chapter explored and described the current reality of sustainable finance, looking at the targets, pathways, standards, frameworks, metrics, methodologies, and attitudes that have shaped the field and the debate. The main purpose was to show that, while sustainability and sustainable finance are moving to the mainstream and driving the transition, this rise to prominence is yet to penetrate the analytical content, framework, and equations of core finance theory and practice.

The discussion revealed that we are still in the early stages of the transformation. Indeed, sustainable finance is coping with variable targets, variable pathways, variable frameworks, variable metrics, variable methodologies, and variable attitudes. Thus, it should not come as a surprise that by October 2021, only 0.5% of $27trn global fund assets were reported to be Paris-aligned, and over 60% of all fund assets were aligned with over 2.75°C (Gonçalves, 2021).

While all the sustainability initiatives discussed in this chapter are truly inspiring, they do not go far enough. They do not transform the dominant value paradigm in finance theory and practice, and ultimately, they leave our mathematics of value and return unchanged.

Our discussion in the previous chapter demonstrated that, in fact, space, as a dimension of analysis and our physical context, has been entirely omitted from the principles and equations of core finance theory and practice. Indeed, our responsibility for space impact is abstracted away. Our discussion in this chapter revealed that sustainable finance, through its standards, frameworks, and metrics, seems to be evolving at the periphery. ESG ratings and integration practices revealed that the approach is not transforming our value models, but instead, affecting specific variables in those models.

This chapter also revealed the necessity to look at sustainable finance as a deeper and broader leap in what we today understand as finance. Indeed, finance must become sustainable, and all finance must be sustainable finance, if we are to overcome the evolutionary challenges we face. We cannot afford a tangential field that falls short of transforming the value paradigm that has, directly or indirectly, contributed to our current predicament.

We must realise and consider that finance theory and practice have a share of responsibility for the climate crisis, and it is through a narrow focus on risk-adjusted returns and the risk-averse mortal investor, and the neglect of space impact, that we have ended up in the current situation.

One of the main arguments put forward in this book is that a consistent and global change across business, industry, and finance requires a radical transformation in our financial value framework. Otherwise, we run the risk of falling into yet another round of partial and opportunistic reductionism that rearranges assets under management, creates a buzz around a few concepts, but ultimately leaves us and the planet on the exact same trajectory we are on.

As I argue in the following chapters, we have a missing principle in finance, the Space Value of Money. The space value of money principle complements the two existing principles in finance and introduces responsibility into the value paradigm and equations of core finance theory and practice, transforming principles as well as equations, making finance sustainable and enabling our transition to a Net Zero future.

References

Adams, C. A, Alhamood, A., He, X., Tian, J., Wang, L., Wang, Y. (2021). The double-materiality concept Application and issues. Global Reporting Initiative. https://www.globalreporting.org/media/jrbntbyv/griwhitepaper-publications.pdf. Accessed 02 February 2022.

Agrawal, A., Hockerts, K. (2018). Impact investing: Review and research agenda. Journal of Small Business & Entrepreneurship, 33(2), 153–181. https://www.tandfonline.com/doi/full/10.1080/08276331.2018.155145. Accessed 02 February 2021.

Ainger, J. (2022). Coal is still raising trillions of dollars despite green shift. BNN Bloomberg. https://www-bnnbloomberg-ca.cdn.ampproject.org/c/s/www.bnnbloomberg.ca/coal-is-still-raising-trillions-of-dollars-despite-green-shift-1.1723066.amp.html. Accessed 15 February 2022.

Avramov, D., Cheng, S., Lioui, A., Tarelli, A. (2021). Sustainable investing with ESG rating uncertainty. Journal of Financial Economics. Forthcoming. https://ssrn.com/abstract=3711218, https://doi.org/10.2139/ssrn.3711218. Accessed 02 February 2022.

Bank of England. (2020). Interim Report and Roadmap for implementing the recommendations of the Taskforce on Climate-related Financial Disclosures. https://www.bankofengland.co.uk/news/2020/november/interim-report-and-roadmap-for-implementing-recommendations-of-taskforce-on-climate-related. Accessed 02 February 2021.

Bank of England. (2021). PRA fines Standard Chartered Bank £46,550,000 for failing to be open and cooperative with the PRA. Press Release. Bank of England. https://www.bankofengland.co.uk/news/2021/december/pra-final-notice-to-standard-chartered-bank-dated-20-december-2021. Accessed 02 February 2022.

Barker, R., Eccles, R. G. (2018) Should FASB and IASB be responsible for setting standards for nonfinancial information? University of Oxford. Said Business School. https://www.sbs.ox.ac.uk/sites/default/files/2018-10/Green%20Paper_0.pdf. Accessed 02 February 2021.

Berg, F., Kölbel, J., Rigobon, R. (2022). Aggregate confusion: The divergence of ESG ratings. https://ssrn.com/abstract=3438533. Accessed 02 February 2022.

Bertl, C. (2016). Environmental finance and impact investing: Status quo and future research. ACRN Oxford Journal of Finance and Risk Perspectives, 5(2), 75–105. http://www.acrn-journals.eu/resources/jofrp0502f.pdf. Accessed 02 February 2021.

Bloomberg. (2021). ESG assets may hit $53 trillion by 2025, a third of global AUM. Bloomberg Intelligence. https://www.bloomberg.com/professional/blog/esg-assets-may-hit-53-trillion-by-2025-a-third-of-global-aum/. Accessed 02 February 2022.

Boffo, R., Marshall, C., Patalano, R. (2020). ESG investing: Environmental pillar scoring and reporting. OECD Paris. https://www.oecd.org/finance/esg-investing-environmental-pillar-scoring-and-reporting.pdf. Accessed 02 February 2022.

Boffo, R., Patalano, R. (2020). ESG investing: Practices, progress and challenges. OECD Paris. https://www.oecd.org/finance/ESG-Investing-Practices-Progress-Challenges.pdf. Accessed 02 February 2022.

Bokern, D. (2022). Reported emission footprints: The challenge is real. MSCI Research. https://www.msci.com/www/blog-posts/reported-emission-footprints/03060866159. Accessed 11 March 2022

Bugg-Levine, A., & Emerson, J. (2011). Impact investing. Jossey-Bass.

CBD. (2021). A new global framework for managing nature through 2030. UN Convention on Biological Diversity. https://www.cbd.int/article/draft-1-global-biodiversity-framework. Accessed 02 February 2022.

CCC. (2019). Net Zero: The UK's contribution to stopping global warming. Climate Change Committee. http://www.theccc.org.uk/wp-content/uploads/2019/05/Net-Zero-The-UKs-contribution-to-stopping-global-warming.pdf. Accessed 12 June 2021.

CDP. (2020). Doubling down: Europe's low-carbon investment opportunity. CDP Europe. https://cdn.cdp.net/cdp-production/cms/reports/documents/000/004/958/original/Doubling_down_Europe's_low_carbon_investment_opportunity.pdf. Accessed 12 December 2021.

CDP-WWF. (2020). Temperature rating methodology. CDP Worldwide and WWF International. https://cdn.cdp.net/cdp-production/comfy/cms/files/files/000/003/741/original/Temperature_scoring_-_beta_methodology.pdf. Accessed 02 November 2021.

CDSB-SASB. (2019). TCFD implementation guide: Using SASB standards and the CDSB framework to enhance climate-related financial disclosures in mainstream reporting. https://www.sasb.org/wp-content/uploads/2019/08/TCFD-Implementation-Guide.pdf. Accessed 02 February 2021.

CISL. (2019). In search of impact measuring the full value of capital. Cambridge Institute for Sustainability Leadership. https://www.cisl.cam.ac.uk/system/files/documents/in-search-of-impact-report-2019.pdf. Accessed 06 June 2021.

CISL. (2020). Measuring business impacts on nature: A framework to support better stewardship of biodiversity in global supply chains. Cambridge Institute for Sustainability Leadership, Cambridge. https://www.cisl.cam.ac.uk/system/files/documents/measuring-business-impacts-on-nature.pdf. Accessed 02 February 2022.

CISL. (2021a). Understanding the climate performance of investment funds. Part 1: The case for universal disclosure of Paris alignment. Cambridge Institute for Sustainability Leadership. https://www.cisl.cam.ac.uk/system/files/documents/understanding-the-climate-performance-of.pdf. Accessed 02 February 2022.

CISL. (2021b). Understanding the climate performance of investment funds. Part 2: A universal temperature score method. Cambridge Institute for Sustainability Leadership. https://www.cisl.cam.ac.uk/download-understanding-climate-performance-investment-funds-part-2. Accessed 02 February 2022.

Christensen, D., Serafeim, G., & Sikochi, A. (2021). Why is corporate virtue in the eye of the beholder? The case of ESG ratings. *The Accounting Review., 97*, 147–175.

CPI. (2021). Global landscape of climate finance 2021. Climate Policy Initiative. https://www.climatepolicyinitiative.org/wp-content/uploads/2021/10/Full-report-Global-Landscape-of-Climate-Finance-2021.pdf. Accessed 02 February 2022.

Dasgupta. P, (2021). The economics of biodiversity: The Dasgupta review. HM Treasury, London. https://assets.publishing.service.gov.uk/government/uploads/system/uploads/attachment_data/file/962785/The_Economics_of_Biodiversity_The_Dasgupta_Review_Full_Report.pdf. Accessed 02 March 2021.

EFRAG. (2021). Proposals for a relevant and dynamic EU sustainability reporting standard-setting. European Financial Reporting Advisory Group. https://www.efrag.org/Activities/2105191406363055/Sustainability-reporting-standards-interim-draft. Accessed 02 February 2022.

Ellerman, A. D., Convery, J. F., & De Perthius, C. (2010). *Pricing carbon: The European Union emissions trading scheme*. Cambridge University Press.

ESA. (2021). ESA's annual space environment report. European Space Agency. https://www.sdo.esoc.esa.int/environment_report/Space_Environment_Report_latest.pdf. Accessed 02 February 2022.

EU. (2018). Action plan: Financing sustainable growth. European Commission. https://eur-lex.europa.eu/legal-content/EN/TXT/?uri=CELEX:52018DC0097. Accessed 02 February 2021.

EU. (2019). European Commission (2019). Guidelines on non-financial reporting: Supplement on reporting climate related information. https://eur-lex.europa.eu/legal-content/EN/TXT/PDF/?uri=CELEX:52019XC0620(01)&from=EN. Accessed 02 February 2022.

EU. (2020a). Sustainable finance taxonomy–Regulation (EU) 2020a/852. European Commission. https://ec.europa.eu/info/law/sustainable-finance-taxonomy-regulation-eu-2020a-852_en. Accessed 02 February 2022.

EU. (2020b). Renewed sustainable finance strategy and implementation of the action plan on financing sustainable growth. European commission. https://ec.europa.eu/info/publications/sustainable-finance-renewed-strategy_en. Accessed 02 February 2022.

EU. (2022). EU taxonomy for sustainable activities. European Commission. https://ec.europa.eu/info/business-economy-euro/banking-and-finance/sustainable-finance/eu-taxonomy-sustainable-activities_en. Accessed 12 February 2022.

EU TEG. (2019). On climate benchmarks and benchmarks' ESG disclosures. European Union Technical Expert Group. https://ec.europa.eu/info/sites/default/files/business_economy_euro/banking_and_finance/documents/190930-sustainable-finance-teg-final-report-climate-benchmarks-and-disclosures_en.pdf. Accessed 06 June 2021.

Fancy, T. (2021). The secret diary of a 'Sustainable Investor'. Medium. https://medium.com/%40sosofancy/the-secret-diary-of-a-sustainable-investor-part-3-3c238cb0dcbf. Accessed 02 February 2022.

FCA. (2021a). NatWest fined £264.8 million for anti-money laundering failures. Press Releases. Financial Conduct Authority. https://www.fca.org.uk/news/press-releases/natwest-fined-264.8million-anti-money-laundering-failures. Accessed 02 February 2022.

FCA. (2021b). FCA fines HSBC Bank plc £63.9 million for deficient transaction monitoring controls. Press Releases. Financial Conduct Authority. https://www.fca.org.uk/news/press-releases/fca-fines-hsbc-bank-plc-deficient-transaction-monitoring-controls. Accessed 02 February 2022.

FCA. (2021c). Credit Suisse fined £147,190,276 (US$200,664,504) and undertakes to the FCA to forgive US$200 million of Mozambican debt. Financial Conduct Authority. https://www.fca.org.uk/news/press-releases/credit-suisse-fined-ps147190276-us200664504-and-undertakes-fca-forgive-us200-million-mozambican-debt. Accessed 02 February 2022.

FCA. (2021d). FCA fines LBGI £90 million for failures in communications for home insurance renewals between 2009 and 2017. Financial Conduct Authority. https://www.fca.org.uk/news/press-releases/fca-fines-lbgi-90-million-failures-communications-home-insurance-renewals-2009-2017. Accessed 02 February 2022.

FFTF. (2011). The Five Capitals—A framework for sustainability. Forum for the Future. https://www.forumforthefuture.org/the-five-capitals. Accessed 02 February 2021.

Friede. G., Busch, T., & Bassen, A. (2015) ESG and financial performance: Aggregated evidence from more than 2000 empirical studies. Journal of Sustainable Finance & Investment, 5(4), 210–233. https://www.tandfonline.com/doi/full/https://doi.org/10.1080/20430795.2015.1118917. Accessed 02 February 2022.

FSB. (2021a). 2021a Status report: Task force on climate-related financial disclosures. Financial Stability Board. https://www.fsb.org/wp-content/uploads/P141021-1.pdf. Accessed 02 February 2022.

FSB. (2021b). Climate-related risks. Financial Stability Board. https://www.fsb.org/work-of-the-fsb/financial-innovation-and-structural-change/climate-related-risks/ Accessed 02 February 2022.

Gibson, R., Krueger, P., Schmidt, P. (2020). ESG rating disagreement and stock returns. Swiss Finance Institute Research Working Paper No. 651/2020. Forthcoming in Financial Analyst Journal. https://papers.ssrn.com/sol3/papers.cfm?abstract_id=3433728. Accessed 02 February 2022.

GIIN. (2016). The business value of impact measurement. Global Impact Investing Network. https://thegiin.org/assets/GIIN_ImpactMeasurementReport_webfile.pdf. Accessed 24 December 2020.

GIIN. (2020). Methodology: For standardizing and comparing impact performance. Global Impact Investing Network. https://thegiin.org/assets/Methodology%20for%20Standardizing%20and%20Comparing%20Impact%20Performance_webfile.pdf. Accessed 02 February 2022.

Gonçalves, P. (2021). Only 0.5% of $27trn global funds assets are Paris-aligned. Investment Week. https://www.investmentweek.co.uk/news/4039267/usd27trn-global-funds-assets-paris-aligned. Accessed 02 February 2022.

GRI. (2021). GRI standards. Global reporting initiative. https://www.globalreporting.org/how-to-use-the-gri-standards/gri-standards-english-language/. Accessed 14 February 2022.

HBS-IWA. (2022). Harvard Business school impact weighted accounts. Harvard Business School. https://www.hbs.edu/impact-weighted-accounts/Pages/default.aspx. Accessed 02 March 2022.

HBS-VBA. (2022). Harvard Business School impact-weighted accounts and value balancing alliance joint statement. https://www.hbs.edu/impact-weighted-accounts/Documents/PI_HBS%20IWA-VBA_Joint%20Statement%20.pdf. Accessed 28 March 2022.

HM Government. (2019). Environmental reporting guidelines: Including streamlined energy and carbon reporting guidance. The National Archives. https://assets.publishing.service.gov.uk/government/uploads/system/uploads/attachment_data/file/850130/Env-reporting-guidance_inc_SECR_31March.pdf. Accessed 06 June 2021.

HSBC-BCG. (2021). Delivering net zero supply chains: The multi-trillion dollar key to beat climate change. Boston Consulting Group and HSBC. https://www.hsbc.com/-/files/hsbc/news-and-insight/2021/pdf/211026-delivering-net-zero-supply-chains.pdf?download=1. Accessed 02 February 2022.

ICAEW. (2021). Knowledge guide to UK accounting standards. Institute of Chartered Accountants in England and Wales. https://www.icaew.com/library/subject-gateways/accounting-standards/knowledge-guide-to-uk-accounting-standards. Accessed 02 February 2022.

IEA. (2017). Energy technology perspectives 2017: Catalysing energy technology transformations. International Energy Agency. https://iea.blob.core.windows.net/assets/a6587f9f-e56c-4b1d-96e4-5a4da78f12fa/Energy_Technology_Perspectives_2017-PDF.pdf. Accessed 02 February 2021.

IEA. (2021). Net Zero by 2050: A roadmap for the global energy sector. International Energy Agency. https://iea.blob.core.windows.net/assets/deebef5d-0c34-4539-9d0c-10b13d840027/NetZeroby2050-ARoadmapfortheGlobalEnergySector_CORR.pdf. Accessed 02 February 2022.

IFRS. (2021). IFRS Foundation announces International Sustainability Standards Board. https://www.ifrs.org/news-and-events/news/2021/11/ifrs-foundation-announces-issb-consolidation-with-cdsb-vrf-publication-of-prototypes/ . Accessed 02 February 2022.

II. (2019). Framework for impact statements: Beta version. The Impact Institute. https://www.impactinstitute.com/wp-content/uploads/2019/04/Framework-for-Impact-Statements-Beta-1.pdf. Accessed 02 February 2021.

IIRC. (2013). Capitals—Background paper for <IR>. The technical task force of the International Integrated Reporting Council. The Technical Collaboration Group. https://www.integratedreporting.org/wp-content/uploads/2013/03/IR-Background-Paper-Capitals.pdf. Accessed 02 February 2021.

IIRC. (2021). International <IR> Framework. The International Integrated Reporting Council. https://www.integratedreporting.org/wp-content/uploads/2021/01/InternationalIntegratedReportingFramework.pdf. Accessed 02 February 2022

IMP. (2016). Impact management norms. Impact Management Project. https://impactmanagementproject.com/impact-management/impact-management-norms/. Accessed 02 February 2021.

Inderst, G., & Stewart, F. (2018). *Incorporating Environmental, Social, and Governance (ESG) factors into fixed income investment*. World Bank Group.

ILB, et al., (2020). The alignment cookbook—A technical review of methodologies assessing a portfolio's alignment with low-carbon trajectories or temperature goal. Institut Louis Bachelier, Institute for Climate Economics, World Wildlife Fund, Ministere De La Transition Ecologique et Solidaire. https://www.institutlouisbachelier.org/wp-content/uploads/2021/03/the-alignment-cookbook-a-technical-review-of-methodologies-assessing-a-portfolios-alignment-with-low-carbon-trajectories-or-temperature-goal.pdf. Accessed 02 February 2022.

IPBES. (2019). The global assessment report on Biodiversity and Ecosystem Services. Intergovernmental Science-Policy Platform on Biodiversity and Ecosystem Services. https://ipbes.net/system/files/2021-06/2020%20IPBES%20GLOBAL%20REPORT%28FIRST%20PART%29_V3_SINGLE.pdf Accessed 02 February 2021.

IPCC. (1988). IPCC history. https://www.ipcc.ch/about/history/. Intergovernmental Panel on Climate Change. Accessed 02 February 2022.

IPCC. (2013). Climate change 2013: The physical science basis. Summary for policymakers. Intergovernmental Panel on Climate Change. https://www.ipcc.ch/site/assets/uploads/2018/03/WG1AR5_SummaryVolume_FINAL.pdf. Accessed 02 February 2021.

IPCC. (2018a). Summary for policymakers. In: Global Warming of 1.5°C. An IPCC Special Report on the impacts of global warming of 1.5°C above pre-industrial levels and related global greenhouse gas emission pathways, in the context of strengthening the global response to the threat of climate change, sustainable development, and efforts to eradicate poverty. IPCC. https://www.ipcc.ch/site/assets/uploads/sites/2/2019/05/SR15_SPM_version_report_LR.pdf. Accessed 02 February 2022.

IPCC. (2018b). Mitigation pathways compatible with 1.5°C in the context of sustainable development. Intergovernmental Panel on Climate Change. https://www.ipcc.ch/site/assets/uploads/2018b/11/sr15_chapter2.pdf. Accessed 02 February 2021.

IPCC. (2021). Climate change 2021: The physical science basis. IPCC. https://www.ipcc.ch/report/ar6/wg1/downloads/report/IPCC_AR6_WGI_SPM_final.pdf. Accessed 02 February 2022.

IPCC. (2022). Climate change 2022: Impacts, adaptation and vulnerability. Summary for policymakers. Intergovernmental Panel on Climate Change. https://report.ipcc.ch/ar6wg2/pdf/IPCC_AR6_WGII_SummaryForPolicymakers.pdf. Accessed 28 February 2022.

Jackson, E. T. (2013). Interrogating the theory of change: Evaluating impact investing where it matters most. Journal of Sustainable Finance & Investment, 3(2), 95–110. https://doi.org/10.1080/20430795.2013.776257. Accessed 02 February 2021.

MEA. (2003). Ecosystems and human well-being: A framework for assessment. Millennium Ecosystem Assessment. Washington, D.C.: Island Press. http://pdf.wri.org/ecosystems_human_wellbeing.pdf. Accessed 02 February 2021.

McKinsey, G. I. (2022). The net-zero transition: What it would cost, what it could bring. Mckensey Global Institute. https://www.mckinsey.com/~/media/mckinsey/business%20functions/sustainability/our%20insights/the%20net%20zero%20transition%20what%20it%20would%20cost%20what%20it%20could%20bring/the%20net-zero%20transition-report-january-2022-final.pdf. Accessed 14 February 2022.

MSCI. (2020). MSCI ESG rating methodology. MSCI ESG Research. https://www.msci.com/documents/1296102/21901542/MSCI+ESG+Ratings+Methodology+-+Exec+Summary+Nov+2020.pdf. Accessed 02 February 2021.

MSCI. (2018). MSCI carbon footprint index ratios methodology. MSCI. https://www.msci.com/documents/1296102/6174917/MSCI+Carbon+Footprint+Index+Ratio+Methodology.pdf/. Accessed 02 February 2022.

NASA. (2021). Space debris and human spacecraft. National. National Aeronautics and Space Administration. https://www.nasa.gov/mission_pages/station/news/orbital_debris.html. Accessed 02 February 2022.

Nusseibeh, S. (2017). The why question. The 300 Club. https://www.the300club.org/wp-content/uploads/2018/10/whitepapers-the-why-question.pdf. Accessed 02 May 2017.

PRI. (2016). A practical guide to ESG integration for equity investing. Principles of Responsible Investing. https://www.unpri.org/download?ac=10. Accessed 02 February 2022.

PRI. (2020). PRI annual report. Principles of Responsible Investment. https://www.unpri.org/download?ac=10948. Accessed 06 June 2021.

Ratsimiveh, K., Hubert, P., Lucas-Leclin, V., Nicolas, E. (2020). ESG scores and beyond Part 1—Factor control: Isolating specific biases in ESG ratings. FTSE Russell. https://content.ftserussell.com/sites/default/files/esg_scores_and_beyond_part_1_final_v02.pdf. Accessed 02 February 2022.

Ratsimiveh, K., Haalebos, R. (2021). ESG scores and beyond Part 2—Contribution of themes to ESG Ratings: A statistical assessment. FTSE Russell. https://content.ftserussell.com/sites/default/files/esg_scores_and_beyond-part_2_final.pdf. Accessed 02 February 2022.

Read, C. (1898). Logic: Deductive and inductive. Simpkin, Marshall, Hamilton, Kent & CO, London. The Project Gutenberg EBook. https://www.gutenberg.org/files/18440/18440-h/18440-h.htm. Accessed 02 February 2021.

Reeder, N., Colantonio, A., Loder, J., Jones, G. R. (2015). Measuring impact in impact investing: An analysis of the predominant strength that is also its greatest weakness. Journal of Sustainable Finance & Investment, 5(3), 136–154. https://www.tandfonline.com/doi/full/https://doi.org/10.1080/20430795.2015.1063977. Accessed 02 February 2021.

Refinitiv. (2021). Environmental, social, and governance scores from Refinitiv. Refintiv. https://www.refinitiv.com/content/dam/marketing/en_us/documents/methodology/refinitiv-esg-scores-methodology.pdf. Accessed 02 February 2022.

SASB. (2020a). SASB implementation supplement: Greenhouse gas emissions and SASB standards. Sustainability Accounting Standards Board. https://www.sasb.org/wp-content/uploads/2020a/10/GHG-Emmissions-100520.pdf. Accessed 02 February 2022.

SASB. (2020b). SASB human capital bulletin. Sustainability Accounting Standards Board. https://www.sasb.org/wp-content/uploads/2020b/12/HumanCapitalBulletin-112320.pdf. Accessed 02 February 2022

SASB. (2020c). Proposed changes to the SASB conceptual framework & rules of procedure—Bases for conclusions and invitation to comment. Sustainability Accounting Standards Board. https://www.sasb.org/wp-content/uploads/2021/07/PCP-package_vF.pdf. Accessed 02 February 2022.

SASB. (2021). SASB standards. Sustainability accounting standards board. https://www.sasb.org/standards/download/. Accessed o2 February 2022.

SASB. (2022). SASB standards & other ESG frameworks. SASB. https://www.sasb.org/about/sasb-and-other-esg-frameworks/. Accessed 03 March 2022.

SEBTi. (2022). Financial sector science-based targets guidance. Science Based Targets Initiative. https://sciencebasedtargets.org/resources/files/Financial-Sector-Science-Based-Targets-Guidance.pdf. Accessed 14 February 2022.

SBTi. (2021). Business ambition for 1.5°C: Responding to the climate crisis. Science Based Targets Initiative. https://sciencebasedtargets.org/resources/files/status-report-Business-Ambition-for-1-5C-campaign.pdf. Accessed 02 February 2022.

SEC. (2022). SEC proposes rules to enhance and standardize climate-related disclosures for investors. US Securities and Exchange Commission. https://www.sec.gov/news/press-release/2022-46. Accessed 21 March 2022.

SEEA. (2014). System of environmental economic accounting 2012—Central framework. United Nations, European Commission, International Monetary Fund, The World Bank, OECD, FAO. https://unstats.un.org/unsd/envaccounting/seeaRev/SEEA_CF_Final_en.pdf. Accessed 02 February 2021.

Serafeim, G., Trinh, K. (2020). A framework for product impact-weighted accounts. Harvard Business School Working Paper 20-076. https://www.hbs.edu/impact-weighted-accounts/Documents/Preliminary-Framework-for-Product-Impact-Weighted-Accounts.pdf. Accessed 02 February 2022.

Sherwood, W. M., & Pollard, J. (2019). Responsible investing: An introduction to environmental, social and governance investments. Routledge.

SIIT. (2014). Measuring impact. The social impact investment taskforce. https://www.gov.uk/government/groups/social-impact-investment-taskforce or https://thegiin.org/research/publication/measuring-impact. Accessed 02 February 2022.

Smolin, L. (2006). *The trouble with physics*. Penguin Books.

Spiess-Knafl, W., & Scheck, B. (2017). Impact investing: Instruments, mechanisms and actors. Palgrave Macmillan.

S&P Global. (2022a). The sustainability yearbook 2022a. S&P Global. https://www.spglobal.com/esg/csa/yearbook/2022a/downloads/spglobal_sustainability_yearbook_2022a.pdf. Accessed 12 February 2022a.

S&P Global. (2022b). S&P Global ESG scores methodology. Sustainable 1. S&P Global. https://www.spglobal.com/esg/documents/sp-global-esg-scores-methodology.pdf. Accessed 28 March 2022b.

TEEB. (2010). The economics of ecosystems and biodiversity: Mainstreaming the economics of nature: A synthesis of the approach, conclusions and recommendations of TEEB. http://www.teebweb.org/wp-content/uploads/Study%20and%20Reports/Reports/Synthesis%20report/TEEB%20Synthesis%20Report%202010.pdf. Accessed 02 February 2021.

TCFD. (2017). Final report: Recommendations of the task force on climate-related financial disclosures. https://assets.bbhub.io/company/sites/60/2020/10/FINAL-2017-TCFD-Report-11052018.pdf. Accessed 02 February 2021.

TCFD. (2021a). Forward looking financial metrics consultation. Task Force on Climate-related Financial Disclosures. https://assets.bbhub.io/company/sites/60/2021/03/Summary-of-Forward-Looking-Financial-Metrics-Consultation.pdf. Accessed 02 February 2022.

TCFD. (2021b). Proposed guidance on climate-related metrics, targets, and transition plans. Task Force on Climate-related Financial Disclosures. https://assets.bbhub.io/company/sites/60/2021b/05/2021b-TCFD-Metrics_Targets_Guidance.pdf. Accessed 01 January 2022.

TCFD-PAT. (2020). Measuring portfolio alignment. https://www.tcfdhub.org/wp-content/uploads/2020/10/PAT-Report-20201109-Final.pdf . Accessed 02 February 2022.

TCFD-PAT. (2021). Measuring portfolio alignment: Technical considerations. https://www.tcfdhub.org/wp-content/uploads/2021/10/PAT_Measuring_Portfolio_Alignment_Technical_Considerations.pdf. Accessed 12 December 2021.

Then, V., Schober, C., Rauscher, O., & Kehl, K. (2017). *Social return on investment analysis*. Palgrave Macmillan.

TNFD. (2021a). Taskforce on nature-related financial disclosures. https://tnfd.global/. Accessed 02 February 2022.

TNFD. (2021b). Nature in scope: A summary of the proposed scope, governance, work plan, communication and resourcing plan of the TNFD. Taskforce on Nature-related Financial Disclosures. https://tnfd.global/wp-content/uploads/2021b/07/TNFD-Nature-in-Scope-2.pdf. Accessed 02 February 2022.

TNFD. (2021c). Proposed technical scope: Recommendations for the TNFD. Taskforce on Nature-related Financial Disclosures. https://tnfd.global/wp-content/uploads/2021c/07/TNFD-%E2%80%93-Technical-Scope-3.pdf. Accessed 02 February 2022.

TNFD. (2022). TNFD releases first beta version of nature-related risk management framework for market consultation. Taskforce on Nature-Related Financial Disclosures. https://tnfd.global/news/tnfd-releases-first-beta-framework/. Accessed 16 March 2022.

UBS-RI. (2022). The responsible investor ESG yearbook 2022. Responsible Investor, UBS AG. https://www.esg-data.com/product-page/esg-yearbook-2022. Accessed 12 February 2022.

UK POST. (2022). POST note: Financial risks of nature loss. The Parliamentary Office of Science and Technology. https://researchbriefings.files.parliament.uk/documents/POST-PN-0667/POST-PN-0667.pdf. Accessed 20 March 2022.

UN. (2016). The sustainable development agenda. United Nations Sustainable Development Goals. https://www.un.org/sustainabledevelopment/development-agenda-retired/. Accessed 12 June 2020.

UN. (2020). The sustainable development goals report. United Nations. https://unstats.un.org/sdgs/report/2020/The-Sustainable-Development-Goals-Report-2020.pdf. Accessed 06 June 2021.

UNEP. (2018). Inclusive wealth report. United Nations Environment Program. https://wedocs.unep.org/bitstream/handle/20.500.11822/26776/Inclusive_Wealth_ES.pdf. Accessed 02 February 2021.

UNEPFI. (2021a). The climate risk landscape: A comprehensive overview of climate risk assessment methodologies. United Nations Environment Program Finance Initiative. https://www.unepfi.org/wordpress/wp-content/uploads/2021a/02/UNEP-FI-The-Climate-Risk-Landscape.pdf. Accessed 02 February 2022.

UNEPFI. (2021b). UNEPFI webpage on principles of responsible banking. https://www.unepfi.org/banking/bankingprinciples/. Accessed 02 February 2022.

UNEPFI. (2022). PRB signatories. UN Environment Program Finance Initiative. https://www.unepfi.org/banking/bankingprinciples/prbsignatories/. Accessed 22 February 2022.

UNFCCC. (2021). COP 26 and the Glasgow Financial Alliance for Net Zero (GFANZ). https://racetozero.unfccc.int/wp-content/uploads/2021/04/GFANZ.pdf. Accessed 02 February 2022.

UNFCCC. (2020). Cities, regions and businesses race to zero emissions. UNFCCC. https://unfccc.int/news/cities-regions-and-businesses-race-to-zero-emissions. Accessed 02 February 2021.

UNFCCC. (2015). Paris agreement. United Nations Framework Convention on Climate Change. https://unfccc.int/sites/default/files/english_paris_agreement.pdf. Accessed 02 December 2020.

US Congress. (2008). Testimony of Dr. Alan Greenspan. House hearing. The financial crisis and the role of federal regulators. https://www.govinfo.gov/content/pkg/CHRG-110hhrg55764/html/CHRG-110hhrg55764.htm. Accessed 02 February 2021.

VBA. (2021a). VBA disclosure concept for material sustainability matters. Value Balancing Alliance. https://www.value-balancing.com/_Resources/Persistent/7/2/a/2/72a28deeed4e259bc414148b2660e631e0dfe3d3/VBA_Disclosure_Concept.pdf. Accessed 02 February 2022.

VBA. (2021b). Methodology—Impact statement. Value Balancing Alliance. https://www.value-balancing.com/_Resources/Persistent/2/6/e/6/26e6d344f3bfa26825244ccfa4a9743f8299e7cf/2021a0210_VBA%20Impact%20Statement_GeneralPaper.pdf. Accessed 12 December 2021a.

VBA. (2021c). Methodology—Impact statement: Extended input-output modelling. Value Balancing Alliance. https://www.value-balancing.com/_Resources/Persistent/0/f/9/1/0f919b194b89a59d3f71bd820da3578045792e2c/2021b0526_VBA%20Impact%20Statement_InputOutput%20Modelling.pdf. Accessed 12 December 2021b.

WRI & WBCSD. (2004). A corporate accounting and reporting standard. World Resources Institute and World Business Council for Sustainable Development. https://ghgprotocol.org/sites/default/files/standards/ghg-protocol-revised.pdf. Accessed 02 February 2021.

WRI & WBCSD. (2011). Corporate Value Chain (Scope 3) accounting and reporting standard. World Resources Institute and World Business Council for Sustainable Development. https://ghgprotocol.org/sites/default/files/standards/Corporate-Value-Chain-Accounting-Reporing-Standard_041613_2.pdf. Accessed 02 February 2021.

4

The Missing Principle: Space Value of Money

> The Earth faces environmental problems right now that threaten the imminent destruction of civilization and the end of the planet as a liveable world. Humanity cannot afford to waste its financial and emotional resources on endless, meaningless quarrels between each group and all others. There must be a sense of globalism in which the world unites to solve the *real* problems that face all groups alike. Can that be done? The question is equivalent to: Can humanity survive?
>
> **Isaac Asimov**, A Memoir, 1994

> When we speak of man, we have a conception of humanity as a whole, and before applying scientific methods to the investigation of his movement we must accept this as a physical fact. But can anyone doubt to-day that all the millions of individuals and all the innumerable types and characters constitute an entity, a unit? Though free to think and act, we are held together, like the stars in the firmament, with ties inseparable.
>
> **Nikola Tesla**, The Problem of Increasing Human Energy, 1900

Undoubtedly, the Net Zero initiative is a necessary strategic objective that will help transform our energy systems, and support business, industry, and finance in achieving their transition to a healthier and more sustainable human economy. However, as discussed in the previous chapter, the current voluntary frameworks of sustainable finance, the associated metrics of alignment, and the strategy of engagement do not provide a rigorous replicable

A. V. Papazian, *The Space Value of Money*,
https://doi.org/10.1057/978-1-137-59489-1_4

method of transition. Temperature and ESG scores are indicators of misalignment, not tools of transformation. Moreover, they do not transform the value paradigm of core finance theory, they do not adjust our equations of value.

In chapter two, we discussed and established that, to date, core finance theory and practice have omitted space, building models that measure the value of money or cash flows based on their risk and time parameters alone, in abstraction. In truth, what the discipline and industry of finance omit is the formal assessment of the value of cash flows relative to space. Our responsibility for space impact has been abstracted away.

A logic of value built vis-à-vis risk and time alone, with the risk-averse investor as only stakeholder, has contributed to the environmental, ecological, and social crises we are facing today. This omission of space and our responsibility for space impact from our analytical framework and equations of value and return can explain why so many need to transition to a responsible future, why the flow of new firms requiring transition is still going strong, and why sectors like coal are still attracting investor attention and funds (Urgewald, 2021, 2022).

To integrate space as a dimension of analysis into core finance theory and practice and establish our responsibility vis-à-vis space as a guiding principle, I propose the introduction of a new and complementary, but as of today still missing, principle: the space value of money.

4.1 The Transition Challenge: Impact & Return

The challenge of the transition is to ensure that investments, assets, projects, and companies do not have a negative space impact—in terms of emissions but also other types of pollution as well. The current value paradigm in finance, built around the time value of money and risk and return, does not prevent investors from investing in opportunities or projects that are in the top left quadrant in Fig. 4.1 (Quadrant 3), where returns are positive, but impact is negative. This is the transition quadrant, and it exists today thanks to the omission of responsibility from our models.

Based on the current value framework taught and applied in finance, the bottom two quadrants in Fig. 4.1 (Quadrants 1 and 2) are automatically dismissed by investors as unattractive.[1] They simply do not provide the required return to the risk-averse return maximising investor. When risk and time adjusted returns are negative, investors have no rationale to invest, even

[1] Note that investor returns refer to expected and required returns, not actual returns, which could of course be negative in different periods due to a variety of business and economic factors.

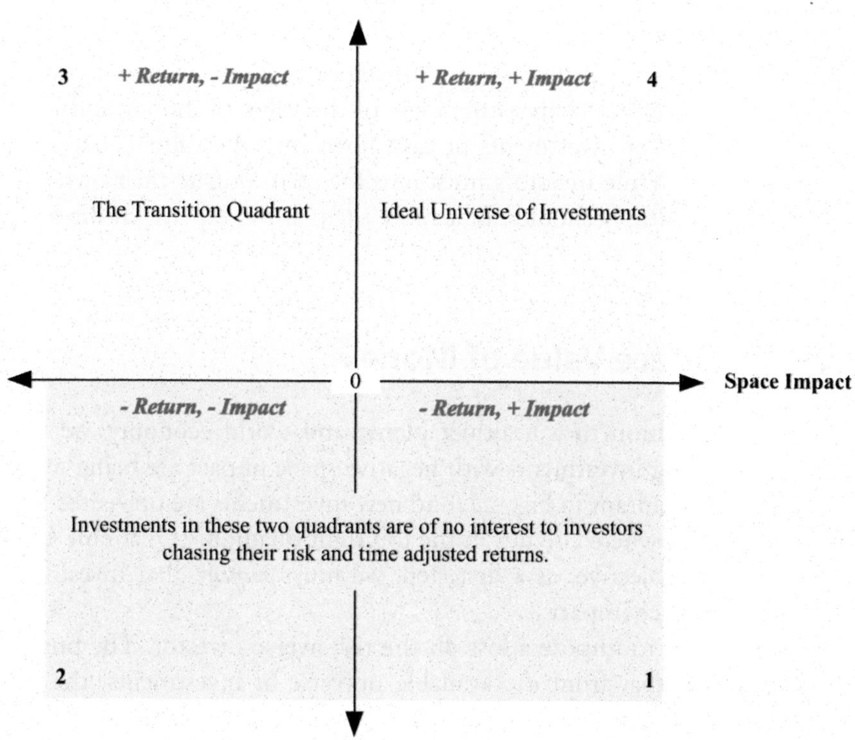

Fig. 4.1 The TRIM: the transition return impact map (*Source* Author)

if the space impact of an investment is positive. Public investments are the exceptions here, which for a variety of reasons may be located in Quadrant 1 and sometimes in Quadrant 2.[2]

Thus, to summarise the transition challenge, to trim negative impact investments we must ensure that investors, who are interested in the top two quadrants, i.e., in positive returns, select those investments that have a positive impact. We must ensure that those investments that have positive returns but negative space impact are never entered into. Indeed, the transition from a finance perspective is about moving investments and businesses into Quadrant 4, where returns and impact are positive.

The simplest and most effective way to ensure investments have no negative space impact is to require that investments have a positive space impact, or, at the very least, have a neutral impact on space. Indeed, the simplest way

[2] Public investments when aimed at providing public goods and services can end up with positive space impacts but negative returns (Quadrant 1), and when they are mismanaged, can end up with negative space impacts and negative returns (Quadrant 2).

is to have a principle that excludes investments with negative space impact, while giving investors their freedom to pursue positive returns.

To achieve such an objective, we can introduce a new principle into finance theory and practice that changes the logic of the value of money and incorporates the impact of investments or cash flows into the value equations of those cash flows. While investors must be allowed to pursue their own positive returns, they simply should not achieve them at the expense of the planet and everyone else.

4.2 The Space Value of Money

If we are to transition to a healthier planet and world economy, we must ensure that existing investments with negative space impact are being moved to the top right quadrant in Fig. 4.1, and new investments are prevented from being initiated anywhere else but in the top right quadrant (Quadrant 4). To achieve such an objective, as a first step, we must *require* that investments have a positive space impact.

The idea is not to impose a loss on the risk-averse investor. The purpose here is to ensure that from the available universe of investments, the risk-averse investor selects those that have a positive impact on space, our physical context. In other words, while pursuing private positive returns, investors should not undertake those opportunities with negative space impact.

The principle I propose, the space value of money, entrenches our responsibility into our financial value framework and requires that a dollar ($1) invested in space has at the very least a dollar's ($1) worth of positive impact on space (Papazian, 2017, 12).[3]

> The Space Value of Money principle requires that a dollar ($1) invested in space has at the very least a dollar's ($1) worth of positive impact on space.

Naturally, this is established as the bottom-line condition, and in truth and for maximum effect, investments must optimise space value per dollar invested. It is common sense that this should be so for all types of investments, for public as well as private investments. Indeed, if we are going to

[3] While I have written about the concept of 'Space Value of Money' in a variety of media articles starting in 2009, I first officially introduced the principle in a paper I presented at Sorbonne University in Paris in 2010. Since then, I have continuously updated and adjusted the framework elements (Papazian, 2011, 2013a, b, c, 2015, 2017, 2020, 2021a, b). The framework and equations presented in this book, while different in terms of depth, scope, and content, have been built upon my past work.

transform and transition, all investments in space should fulfil the space value of money principle before being taken on. Space value of money complements time value of money and risk and return and changes our optimisation target: while we are maximising our returns and minimising our risks, we *must* also optimise our space impact. The principle entrenches responsibility into our models and requires that our mathematical expressions of value reflect this fact. I expand on the metrics of space value in the next chapter.

Revisiting the Transition Return Impact Map with the space value of money principle, Fig. 4.2, we can see how the principle allows us to eliminate negative impact investments (Quadrants 3 and 2), without constraining the pursuit of positive investor returns. Indeed, the principle supports the transition and ensures that new investments are initiated in the positive impact/return quadrant (Quadrant 4), the ideal universe of investments.

Given our mid-century Net Zero target and the Race to Zero, the space value of money principle, and the associated metrics discussed in later chapters, can provide the type of change needed to entrench sustainability into

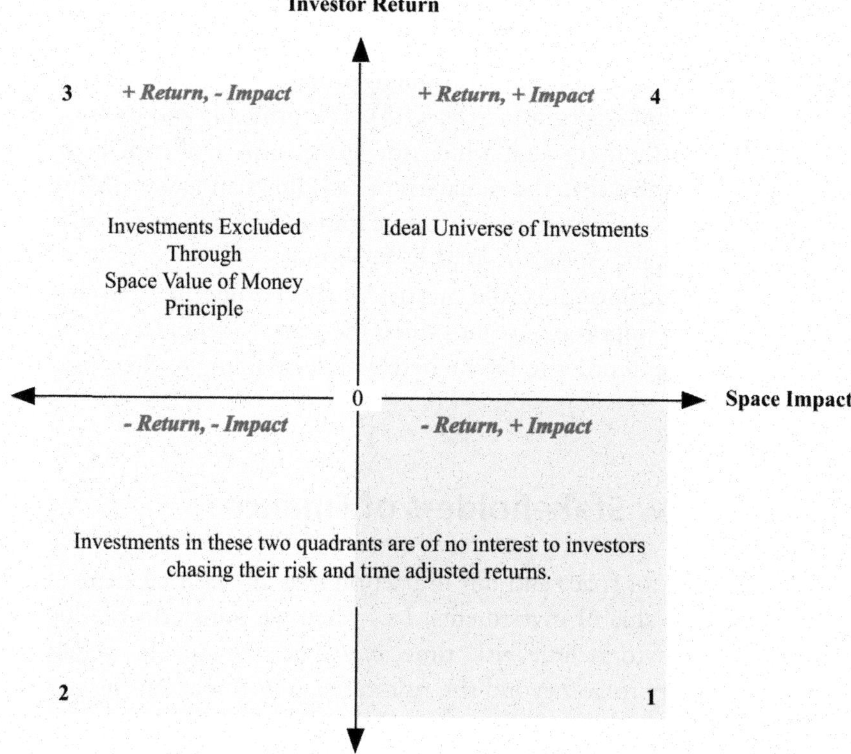

Fig. 4.2 The TRIM and the space value of money (*Source* Author)

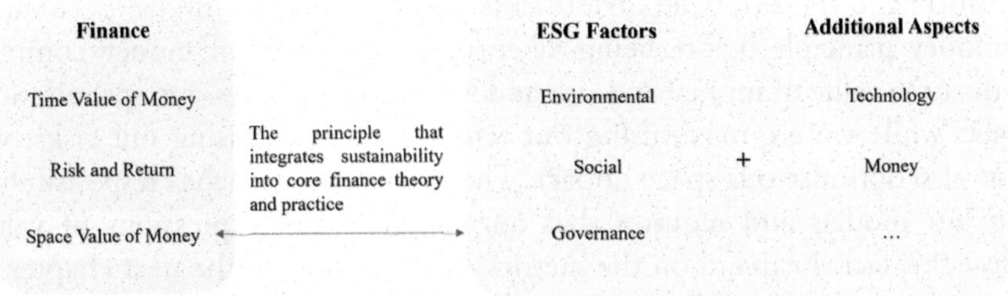

Fig. 4.3 ESG factors and the Space Value of Money (*Source* Author)

finance. If we are to achieve our objectives, the risk-averse investor cannot remain the only stakeholder in our core financial value models and equations, while sustainable finance expands on its margins with a voluntary framework focused on engagement, with metrics and scores that act as indicators of misalignment rather than tools of transformation. This is the core challenge of sustainable finance, or finance, and given the state of the world and our ecosystem, the magnitude of the challenge and the solutions required will undoubtedly involve a paradigm shift within the discipline.

The space value of money principle can be considered the theoretical link between core finance theory and the many different trends in sustainable finance including ESG integration (Fig. 4.3). The principle and its associated metrics offer a method through which the space impact of cash flows and assets can be integrated into the valuation of cash flows and assets. Moreover, given the broader conceptualisation of space impact, the framework allows us to consider all relevant aspects beyond E, S, and G.

I will discuss and introduce the metrics of space value in later chapters. At this stage, it is important to introduce the new stakeholders of finance theory and practice in order to have a better understanding of their priorities in space.

4.3 The New Stakeholders of Finance

When we account for space and our impact on it as an essential element and ingredient of the value of investments, i.e., when we transform the logic of the value of money to include, risk, time, *and* space parameters, we face the theoretical necessity to go beyond the risk-averse return maximising investor as sole stakeholder, to identify the new stakeholders that come into play and discuss their preferences and specificities.

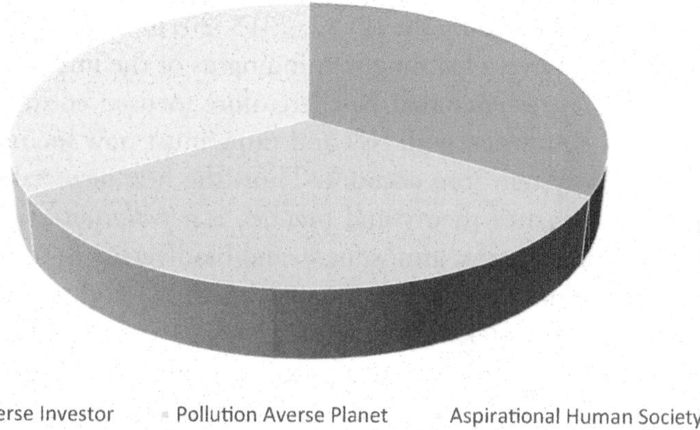

Fig. 4.4 The three new stakeholders of finance (*Source* Author)

Outside of the risk-averse investor, considered through time and risk parameters, there are two main stakeholders that we must identify in space. The first most obvious stakeholder is our home planet, as an ecosystem that ensures our survival and provides a platform through and upon which we express and experience our own productivity. The second equally obvious stakeholder is humanity as a whole—not earning a monetary return but sharing the impact of investments in space.

Figure 4.4 depicts the two new stakeholders of finance theory and practice alongside the risk-averse investor. In the following sections, I explore the key features of these two new stakeholders that must be introduced into our analysis and equations of value.

4.3.1 A Pollution-Averse Planet

When conventional finance theory builds its models, it does so having the individual investor in mind, characterised mainly by a risk-averse nature. What would be the key feature of our planet when introduced as a new stakeholder in finance?

The planet effectively provides everything we need to survive and thrive. Human productivity on Earth has been shaped by our own evolution, driven by our own imagination and technological advances. Whatever the reasons behind specific industrial processes, our understanding of energy and the many sources available to us, and the rationale behind our financial and monetary structures, as of today, we have been continuously usurping the environment that feeds us, destroying its biodiversity, ruining its air, polluting its oceans, etc.

The evidence provided by the IPCC (2013, 2018, 2021, 2022) and IPBES (2019) is clear and overwhelming, reminding us of the impact we are having on our planet. The science that has heretofore focused entirely on the risk-averse investor, concerned with risk and time, must now include our planet, and take our ecosystem into account. Thus, the first new stakeholder to be introduced into finance theory and practice is a *pollution-averse planet* that values its air, lands, oceans, atmosphere, and biodiversity as much as the risk-averse investor values her/his/their financial mobility and security.

In truth, this new stakeholder is our physical context, from the atomic fabric of matter to interstellar space, and not just the planet as such (See Fig. 4.5). This is so simply because we have already extended our reach far

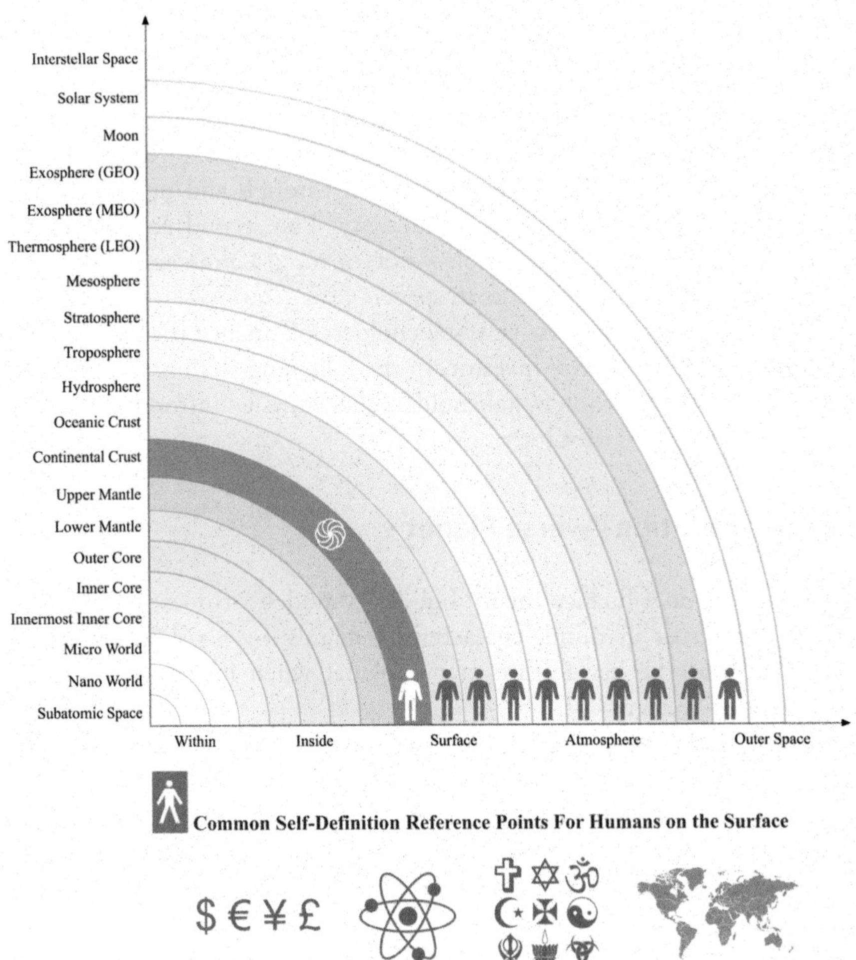

Fig. 4.5 Space layers and some example identity reference points (*Source* Author)

beyond the planet with a human crew in the International Space Station, rovers roaming around Mars, and Voyager 1 in interstellar space. Similarly, we are already active in the atomic and subatomic worlds, from genetic engineering to nanotechnology, we have been manipulating matter on many levels.

When we discuss the pollution-averse planet, we must realise that we are not talking about emissions alone. All types of pollution and waste are included here. Respecting the natural ecosystem that feeds and sustains us is critical, and it is quite absurd that somehow the mortal risk-averse investor has been given priority to date. The reasons behind this abstraction rest with our financial value framework, and as I will show in chapter eight, in the monetary architecture we have built on the time and risk focused value framework.

It is almost a Kafkaesque infliction on the healthy, balanced, and respectful human mind that it must convince the unconscious and short-sighted exploitation and destruction of our only ecosystem to cease and desist. The science behind climate change and the climate scientists who have taken on what Mann (2021) calls the new climate war are a testimony to this absurdity.

While powerful interests are without a doubt responsible for our current state of affairs, and it is only natural that they fight back to preserve the structures and value framework that has given them the power they hold, it is also entirely true that the power to change the science of finance, and trigger the necessary transformations in practice, is in the minds of millions of students, analysts, scholars, and investors spread around the planet.

We must imagine and reason the necessary transformations in our minds to be able to translate them into improved principles, equations, institutions, and digital tools. The first step is to formally introduce our planet as a lawful and rightful stakeholder in our financial value models.

4.3.2 An Aspirational, Eternal, and Growing Human Society

I refer to the risk-averse investor as a mortal investor because it helps define the risk and time focused nature of the investor. An immortal being, who needs no medicine and cannot die in the cold, has no, or less, concern for risk, and time. Most probably, however, though we cannot assert this with any certainty, the search for meaning in an immortal body will be equally intense.

While finance focuses on the individual mortal human in the chain, it also omits any formal consideration of the whole, i.e., humanity, as a species and as a collective reality that lives and survives on planet Earth.

The next stakeholder is humanity, including every human being on this planet, including future generations being born today, or those who will be born sometime in the near or far future. The dead are also part of the emotional, spiritual and intellectual heritage of humanity, but from a physical point of view, they are part of the previous stakeholder.

If the individual investor is risk-averse, what key features would we use to describe the preferences of this new stakeholder? Humanity is a far more complex organism and its preferences cannot possible be reduced to one such feature, like 'risk-averse'. Indeed, the following discussion does not claim to be exhaustive or comprehensive. It is an attempt to crystallise the key features that must be taken into account when introducing humanity into the value equations of finance theory and practice.

Humanity is still a fragmented entity due mainly to the diversity of self-interpretations, on individual and collective levels. While as a species we are one, our identities vary. In truth, the foundations of identity vary as well. A quick snapshot of humanity today would reveal that humans use a variety of different reference points when building their self-identity, or when interpreting themselves, their associations, and their existence.

Some interpret themselves and their life in relation to the political map, by location, region, or by the nation or state they are born into or feel they belong to. Others interpret themselves and their life in relation to the content of a specific book that they consider to be holy, which shows them the way, and provides guidelines through associated institutions. Others interpret themselves and their lives in relation to a subject matter and their curiosity to understand it and contribute to it—built around a collection of ideas, observations, and facts covering potentially anything in the universe. Others define themselves through their material and monetary acquisitions and status. Others define and understand themselves as mass and energy made of atoms, particles, and waves. Others define and understand themselves in relation to stars and constellations.

Indeed, if we were to look at humanity in its physical context, stretching from atoms to the planet and beyond, we can comfortably recognise that the spatial horizon varies from person to person. While the next chapter will dive deeper into space and the many layers of space we have come to affect or interact with, the point here is to note that humans do not use the same starting point of reference for their self-interpretation. Thus, when we speak of humanity, we must understand and respect the mosaic of interpretations, from a spatial perspective, as well as content perspective.

Figure 4.5 depicts our physical context stretching from within matter, to inside the planet, the surface, the atmosphere, and outer space. It also depicts those layers of space where we have physically touched in our human form. While we study and have touched all the identified layers of space through technology, we have physically accessed the ones identified. The figure also identifies four commonly observed self-definition reference points on the surface of the planet (Continental Crust).

Naturally, the reference points I mention here are neither mutually exclusive nor comprehensive and most often overlap to a certain extent. The point here is that our freedom of self-interpretation is the most important freedom from which all other freedoms and virtues and values are born. We have the freedom to interpret ourselves, our lives, and our meanings in the universe. As such, this stakeholder requires and deserves freedom and respect whatever its self-interpretation and/or identity. This, I believe, is one of the most important features of our new stakeholder.

Whatever the number and types of identity we encounter within humanity, across all modalities of self-interpretation, our biology is identical, our chemistry is identical, and as a carbon-based life form, we live and die, and have been doing so since the beginning of our time, on a planet in space, spinning on itself at around 1000 miles per hour, orbiting the sun at a 92.96 million miles distance, with a 67,000 mph speed.

Day and night, billions of humans living across all continents (and in the International Space Station), dream, work, produce, and aspire for a bright or brighter future—whether we define this as desire to achieve justice, fairness, equality, or ambition to reach the moon and the stars, ambition to invent a cure for cancer, to succeed, achieve a healthy standard of living, a happy home, etc. Whatever our definition of a bright or brighter future, we all want it for ourselves and our children. This is the next feature of our new stakeholder.

Humanity is curious, imaginative, and determined to pursue its aspirations and ready to invent all that it needs to invent to achieve them. The last one hundred or so years are a testimony of our capabilities to achieve anything we put our mind to. Indeed, we are an aspirational human society, with the imagination and will to pursue our diverse ambitions. Thus, this stakeholder requires and deserves fair opportunities to grow and realise its aspirations, individually and collectively, without prejudice or discrimination.

This stakeholder is also growing continuously and has been doing so for the last many centuries. While annual growth rates have been declining, actual totals are still increasing, and projected to reach 10.9 billion by the end of the century (Roser, 2019). Currently comprised of more than 7.9 billion souls (UNPF, 2021), humanity is an expanding entity that can, in theory and in reality, continue to live and secure its survival through procreation. Thus, unless we self-destruct due to self-inflicted short-sightedness and deep misinterpretations of our reality and place in the universe, humanity may continue to live and expand through the universe eternally.

As such, our new stakeholder is free to self-define, aspires to achieve a brighter future for itself and its future generations, it is growing continuously, and short of any self-inflicted cataclysm or extinction level event (ELE), is eternal. Which means that successfully fulfilling its evolutionary ambitions and the sustainable expansion of its productivity on earth and beyond are key targets.

4.4 The Risk and Time Features of Our Evolutionary Challenges

In finance, from the perspective of the mortal risk-averse investor, time is defined as calendar time. Usually, given the risk-averse mortal nature of the investor, the time horizon is short. In some cases, it can stretch for a few decades, crossing a generation or two. When we look at time from the perspective of an eternal species, horizons and priorities change, even the very nature of risk changes. Think about all our evolutionary challenges, those created by our own short-sightedness, like climate change and environmental degradation, and those due to the vastness of space and the mysteries of life. None of these challenges are risk-free. In fact, they all have incalculable risks.

All evolutionary investments promise intergenerational returns that could only be materialised in a distant future, carrying immense levels of risks in the present. With a time horizon that stretches beyond our lifetime, and risks that cannot objectively be summarised through a discount rate, evolutionary challenges are in a blind spot.

Indeed, our financial value framework to date is directly and indirectly responsible for the business and industrial practices that have ruined our air, our oceans, our lands, and even outer space. If finance were to value the impact of cash flows and assets as an integral part of the value of investments, no project or asset or investment could ever get away with a negative impact on space.

But the challenge is deeper and even more debilitating than just pollution and degradation. Our current financial models hinder human evolution because the risk and time value principles cannot cope with evolutionary investments, where risks are off the charts and returns are too distant.

Our risk and time-based value paradigm entrenches a negative bias towards distant payments in the future due to the discounting methodology used. The further away in time is an expected cash flow, the lower its present value given the discount rate chosen to reflect its risk levels. In other words, our value paradigm discriminates against evolutionary ideas and investments, given the risk and return and time value principles.

The current value paradigm of core finance theory and practice is architecturally designed to leave our evolutionary investments in a blind spot, as the principles by their very nature discriminate against highly risky projects with distant returns. While this is not an intentional bias, it is nevertheless the consequence of the main principles of value adopted and used in the field. Indeed, this is why the introduction of the space value of money principle and associated metrics is relevant and necessary. The space value of money framework allows us the opportunity to rectify this imbalance and provides, as the following chapters will show, a methodology through which our space impact in the present can be valued, and when necessary, compounded into the future.

4.4.1 Short-Termism and Quarterly Capitalism

The proposition discussed in the previous section, while in agreement with the short-termism argument that has attracted significant attention recently, goes further and addresses a deeper architectural deficiency born from the risk/time focus of our financial value paradigm.

Indeed, there is an entire literature that considers and measures the undesired impact of short-termism, or the short-term focus on quarterly profits, which affects research and development, innovation, and also investment and financial stability (Barton, 2011; Haldane & Davies, 2011; Stein, 1989). Haldane and Davies in their speech titled the 'The Short Long' (2011) at the 29th Colloquium of the Société Universitaire Européene de Recherches Financières addressing the overall topic of 'New Paradigms in Money and Finance' state the following:

> Our evidence suggests short-termism is both statistically and economically significant in capital markets. It appears also to be rising. In the UK and US, cash-flows 5 years ahead are discounted at rates more appropriate 8 or more years hence; 10 year ahead cash-flows are valued as if 16 or more years ahead;

> and cash-flows more than 30 years ahead are scarcely valued at all. The long is short. (Haldane & Davies, 2011, 1)

The short-termism research and evidence reveal that even within a time-based model which discriminates against much longer-term evolutionary projects by default, attention spans are shrinking, both in terms of time horizon, but also risk. As Haldane and Davies reveal, the focus is moving to shorter time horizons, and discount rates being applied to 5-year cash flows are more appropriate to 8+ years, and cash flows 30 years ahead are 'scarcely valued at all'.

The impact of short-termism stretches far beyond research and development, innovation, and financial stability. It also feeds a culture of productivity and profit maximisation within which responsibility and environmental impact have a very tangential role. What Barton (2011) describes as 'quarterly capitalism' is a central challenge, and as he suggests, shifting to long term capitalism must involve 'rewiring the fundamental ways we govern, manage, and lead corporations… [and] how we view business's value and its role in society' (Barton, 2011).

As critical as it is to address this short-termism in our corporate and financial reality, we must realise that it is driven by an incentive structure shaped by an entirely time and risk focused financial value paradigm. A paradigm that leaves our evolutionary investments in a blind spot due to the internal biases of the principles it is built on.

Within this time/risk focused framework, the accelerated shrinking of the horizon, or short-termism and quarterly capitalism, even if addressed, leaves us no closer to rectifying the inherent biases towards investments with incalculable or very high risks and very distant returns, i.e., all our evolutionary investments addressing our many evolutionary challenges.

Indeed, Carney's (2015) description of climate change as 'the Tragedy of the Horizon' reveals this short and shrinking time horizon phenomenon from a macro perspective.

> We don't need an army of actuaries to tell us that the catastrophic impacts of climate change will be felt beyond the traditional horizons of most actors—imposing a cost on future generations that the current generation has no direct incentive to fix.
>
> That means beyond: the business cycle; the political cycle; and the horizon of technocratic authorities, like central banks, who are bound by their mandates. The horizon for monetary policy extends out to 2–3 years. For financial stability it is a bit longer, but typically only to the outer boundaries of the credit cycle—about a decade. In other words, once climate change

becomes a defining issue for financial stability, it may already be too late. (Carney, 2015, 3)

As important as the reforms identified by Barton (2011) and the short and shrinking horizon challenges discussed by Haldane and Davies (2011) and Carney (2015) are, the point here is that a shrinking time/risk horizon within a time/risk framework can only be addressed by integrating space into the value paradigm, and not by stretching time horizons as such.

In truth, we need to introduce space into our analytical framework, establish our responsibility, and in parallel to discounting future expected cash flows to the present, we need to start compounding our impact in the present into the future. More on this in later chapters.

4.4.2 Funding Evolutionary Challenges

The best way to contextualise the discussion is to visit the question that is most pressing today: how are we going to fund the immense costs of transition to Net Zero on a global scale? This is still very much a question of debate and speculation.

The Climate Policy Initiative (CPI, 2014) states that 'based on IEA projections, we estimate that approximately $34 trillion of investment in energy infrastructure will be required from 2015 through 2030 regardless of climate goals. We estimate that the 2DS scenario requires approximately $11 trillion of additional investment over the same time period'. More recently, the Energy Transition Commission (ETC) announced that we will need between $1 trillion–$2 trillion a year to achieve this by mid-century (Chestney, 2020). In a recent article published on the eve of COP26 in the Financial Times, Mark Carney, former governor of the Bank of England and UN Special Envoy on Climate Action and Finance, argues that the $100 trillion climate challenge must be met by the finance industry (FT, 2021; McNish & Hoffman, 2021).

A recent HSBC and Boston Consulting Group report assesses the transition costs of entire supply chains across industries at $100 trillion and they state that '[t]he longer this investment is delayed, the greater the amount required, and the more extreme the damage from storms, fires, droughts, and ecosystem harm (to cite only a few forms of damage) in the meantime' (HSBC-BCG, 2021, 3). A recent report by the McKinsey Global Institute quantifies the costs at $275 trillion and identifies the key challenges we face when taking action (McKinsey GI, 2022, 12):

> [F]irst, the scale and pace of the step-up in spending needed on physical assets, given that entire energy and land-use systems evolved over a century or two and would need to be transformed over the next 30 years; second, the collective and global action required, particularly as the burdens of the transition would not be evenly felt; third, the nearterm shifts needed for longer-term benefits; fourth, the shifts needed in business practices and lifestyles that have evolved over decades; and fifth, the central role of energy in all economic activity, which means that transformation would need to be carefully managed.

To shed some further light on the matter, in 2009, at COP15 of the UN Climate Change Conference (UNFCCC, 2015, 2021), parties agreed to raise $100 billion a year for climate finance for developing countries by 2020, and this target has been missed every year since 2013 (HOCL, 2021).

The reason the funding for a global transition to Net Zero is still a matter of debate and discussion is because our value framework, markets, and instruments in finance do not provide immediate answers regarding the source and mechanism of deployment for such funding. With a risk and time focused value paradigm that ignores space and serves the mortal risk-averse investor, finance theory and practice have no ready answer to such a challenge.

As discussed in the earlier sections, this is so because evolutionary investments, due to their time and risk profiles, are conceptually and mathematically challenging for the field. A framework that discriminates against evolutionary investments due to its inherent biases vis-à-vis risk and time will fall short of meeting this challenge without fundamental reforms. This is also linked to our monetary architecture, and it is discussed in detail in chapter eight.

Moreover, the climate crisis is not our only evolutionary challenge. Jeffrey Sachs (2006) in his book the End of Poverty conjectured that it would take $175 billion per year for 20 years to eradicate poverty, that is $3.5 trillion over 20 years. While he proposes aid as a possible solution, given the relative size of the amount when compared to the income of rich countries, the fact that such solutions are not forthcoming is another testimony to the fact, in support of the argument being made here.

Evolutionary challenges that need financing are not all current crises. Evolutionary challenges that are not crises yet, but deserve investment, are many. One such frontier is outer space exploration. We are a carbon-based life form on a planet orbiting a star that will probably die out in 5 billion years (NASA, 2021). Assuming we do not destroy ourselves and our planet much before, we have every reason to explore this universe, and we must. Indeed, given the life cycle of our solar system and our sun, many have argued that we must establish ourselves as an interplanetary species (Musk, 2017; Hawking, 2008; Sinclair and Lee et al., 2012).

Naturally, while none of this should be done at the expense of our home planet and the well-being of those living on it, we should clearly recognise that outer space exploration has had an invaluably positive impact on planetary sustainability and on our own technological advancement in many fields and industries. Indeed, besides the technological advancements, outer space exploration has had a very deep and irreversible impact on human consciousness as well. Since the 1968 Earthrise picture, dubbed 'the most influential environmental photograph ever taken' by Galen Rowell, we have come to see our own home planet from the outside, the 'overview effect' that has radically changed the life of astronauts and our perceptions and interpretations of our home planet (White, 1987).

NASA, William Anders, Apollo 8, 24/12/1968 (NASA 2022)
Source https://www.hq.nasa.gov/office/pao/History/alsj/a410/AS8-14-2383HR.jpg

In chapter eight, I explore and discuss the innovative funding solutions that will become possible once we have transformed our value paradigm, adopted the space value of money principle, and introduced the many metrics of space impact discussed in the following chapters.

4.5 Conclusion

The space value of money principle is proposed as the missing principle in finance, complementing the time value of money and risk and return—two principles that have left our evolutionary investments in a blind spot due to their inbuilt bias against highly risky and distant returns.

Indeed, the time and risk focus of finance theory to date, serving the mortal risk-averse investor, could be considered a theoretical failure of cosmic magnitude. A world on the edge of irreversible ecological collapse is the by-product of a financial value framework that has omitted space and our responsibility of impact in it and on it.

By introducing space into our financial value framework, as an analytical dimension and as our physical context stretching from inside matter to interstellar space and beyond, we are now required to adjust our models such that the impact of investments on space are accounted for in our equations of value and return. When we introduce space into the analysis, and we commit to considering the space impact of cash flows an integral element of the value of cash flows, we go beyond the risk-averse investor and account for two new stakeholders, a pollution-averse planet and an aspirational eternal human society.

The principle of space value of money states that a dollar invested in space must have at the very least a dollar's worth of positive impact on space. By doing so, it establishes the bottom threshold of acceptable investments and opportunities. The principle entrenches spatial responsibility into our financial value framework and ensures that no investment with a negative space impact is considered investable ever again. It also transforms our optimisation target: we must optimise space impact as we minimise risks and maximise returns.

As such, finance theory and practice must be adjusted not just for ESG. While the buzz around ESG is understandable given the lack of theoretical clarity around the subject, our challenge is a much deeper issue with responsibility in our financial models, with principles and equations that are structured to serve a mortal human in the chain, rather than the eternal species and its evolutionary ambitions.

The introduction of space as an analytical dimension and the entrenchment of our responsibility in space through the principle of space value of money can offer us the opportunity to truly address our evolutionary challenges.

References

Ainger, J. (2022). Coal is still raising trillions of dollars despite green shift. BNN Bloomberg. https://www-bnnbloomberg-ca.cdn.ampproject.org/c/s/www.bnnbloomberg.ca/coal-is-still-raising-trillions-of-dollars-despite-green-shift-1.1723066.amp.html. Accessed 15 February 2022.

Asimov, I. (1994). *A memoir*. Bantam Books.

Barton, D. (2011). Capitalism for the long term. Harvard Business Review. https://hbr.org/2011/03/capitalism-for-the-long-term. Accessed 02 February 2022.

Carney, M. (2015). Breaking the tragedy of the horizon—Climate change and financial stability. Bank of International Settlements FSB. Available at https://www.bis.org/review/r151009a.pdf. Accessed 26 December 2018.

Chestney, N. (2020). Global net zero emissions goal would require $1–2 trillion a year investment—study. Reuters News. https://www.reuters.com/article/uk-energy-transition-idUKKBN2670OA. Accessed 02 January 2022.

CPI. (2014). Moving to a low-carbon economy: The financial impact of the low carbon transition. Climate Policy Initiative. https://climatepolicyinitiative.org/wp-content/uploads/2014/10/Moving-to-a-Low-Carbon-Economy-The-Financial-Impact-of-the-Low-Carbon-Transition.pdf. Accessed 02 February 2022.

FT. (2021). Mark Carney: The world of finance will be judged on the $100tn climate challenge. Financial Times. https://www.ft.com/content/d9e4ebb9-f212-406a-90d5-73b4276539e6. Accessed 12 December 2021.

Haldane, A. G., & Davies, R. (2011). The short long. Bank of England. 29th Société Universitaire Européene de Recherches Financières Colloquium: New Paradigms in Money and Finance? https://www.bankofengland.co.uk/-/media/boe/files/speech/2011/the-short-long-speech-by-andrew-haldane.pdf. Accessed 02 February 2022.

Hawking, S. (2008). Why we should go into space. National Space Society. https://space.nss.org/stephen-hawking-why-we-should-go-into-space-video/. Accessed 02 February 2022.

HOCL. (2021). COP26: Delivering on $100 billion climate finance. House of Commons Library. https://commonslibrary.parliament.uk/cop26-delivering-on-100-billion-climate-finance/. Accessed 12 January 2022.

HSBC-BCG. (2021). Delivering net zero supply chains: The multi-trillion dollar key to beat climate change. Boston Consulting Group and HSBC. https://www.hsbc.com/-/files/hsbc/news-and-insight/2021/pdf/211026-delivering-net-zero-supply-chains.pdf?download=1. Accessed 02 February 2022.

IPBES. (2019). The global assessment report on Biodiversity and Ecosystem Services. Intergovernmental Science-Policy Platform on Biodiversity and Ecosystem Services. https://ipbes.net/system/files/2021-06/2020%20IPBES%20GLOBAL%20REPORT%28FIRST%20PART%29_V3_SINGLE.pdf Accessed 02 February 2021.

IPCC. (2013). Climate change 2013: The physical science basis. Summary for policymakers. Intergovernmental Panel on Climate Change. https://www.ipcc.ch/site/assets/uploads/2018/03/WG1AR5_SummaryVolume_FINAL.pdf. Accessed 02 February 2021.

IPCC. (2018). Summary for policymakers. In Global warming of 1.5 °C. An IPCC Special Report on the impacts of global warming of 1.5 °C above pre-industrial levels and related global greenhouse gas emission pathways, in the context of strengthening the global response to the threat of climate change, sustainable development, and efforts to eradicate poverty. IPCC. https://www.ipcc.ch/site/assets/uploads/sites/2/2019/05/SR15_SPM_version_report_LR.pdf. Accessed 02 February 2022.

IPCC. (2021). Climate change 2021: The physical science basis. IPCC. https://www.ipcc.ch/report/ar6/wg1/downloads/report/IPCC_AR6_WGI_SPM_final.pdf. Accessed 02 February 2022.

IPCC. (2022). Climate change 2022: Impacts, adaptation and vulnerability. Summary for Policymakers. Intergovernmental Panel on Climate Change. https://report.ipcc.ch/ar6wg2/pdf/IPCC_AR6_WGII_SummaryForPolicymakers.pdf. Accessed 28 February 2022.

Mann, M. E. (2021). *The new climate war: The fight to take back our planet*. Public Affairs Books.

McKinsey GI. (2022). The net-zero transition: What it would cost, what it could bring. Mckensey Global Institute. https://www.mckinsey.com/~/media/mckinsey/business%20functions/sustainability/our%20insights/the%20net%20zero%20transition%20what%20it%20would%20cost%20what%20it%20could%20bring/the%20net-zero%20transition-report-january-2022-final.pdf. Accessed 14 February 2022.

McNish, J., & Hoffman, L. (2021). Mark Carney, Ex-Banker, wants banks to pay for climate change. *The Wall Street Journal*. https://www.wsj.com/articles/mark-carney-ex-banker-wants-banks-to-pay-for-climate-change-11635519625. Accessed 12 January 2022.

NASA. (2021). Our Sun. National Aeronautics and Space Administration. https://solarsystem.nasa.gov/solar-system/sun/in-depth/. Accessed 02 February 2022.

NASA. (2022). Earthrise picture. National Aeronautics and Space Administration. https://www.hq.nasa.gov/office/pao/History/alsj/a410/AS8-14-2383HR.jpg. Accessed 02 February 2022.

Papazian, A. V. (2011). A product that can save a system: Public capitalisation notes. In Collected seminar papers (2011–2012) (Vol. 1). Chair for Ethics and Financial Norms. Sorbonne University.

Papazian, A. V. (2013a). Space exploration and money mechanics: An evolutionary challenge. International Space Development Hub. https://papers.ssrn.com/sol3/papers.cfm?abstract_id=2388010. Accessed 06 June 2021.

Papazian, A. V. (2013b). Our financial imagination and the Cosmos. Cambridge Judge Business School. University of Cambridge. https://www.jbs.cam.ac.uk/insight/2013b/our-financial-imagination-and-the-cosmos/. Accessed 12 December 2021.

Papazian, A. V. (2013c). Economics and finance: The frontlines of galactic evolution. *Principium* (7). 8–9. https://i4is.org/wp-content/uploads/2017/01/Principium_7_NovDec_2013.pdf. Accessed 12 February 2022.

Papazian, A. V. (2015). Value of money in spacetime. The International Banker. https://internationalbanker.com/finance/value-of-money-in-spacetime/. Accessed 02 February 2021.

Papazian, A. V. (2017). The space value of money. review of financial markets (Vol. 12, pp. 11–13). Chartered Institute for Securities and Investment. Available at: http://www.cisi.org/bookmark/web9/common/library/files/sironline/RFMJan17.pdf. Accessed 16 December 2020.

Papazian, A. V. (2020). An algorithm for responsible prosperity: A new value paradigm. Available at: https://ssrn.com/abstract=3633763. Accessed 12 February 2022.

Papazian, A. V. (2021a). Towards a general theory of climate finance. Available at: https://ssrn.com/abstract=3797258. Accessed 12 February 2022.

Papazian, A. V. (2021b). Sustainable finance and the space value of money. Cambridge Judge Business School. University of Cambridge. https://www.jbs.cam.ac.uk/insight/2021/sustainable-finance-and-the-space-value-of-money/. Accessed 02 February 2022.

Roser, M. (2019). Future population growth. Our World in Data. https://ourworldindata.org/future-population-growth. Accessed 12 January 2022.

Sachs, J. (2006). *The end of poverty*. Penguin Books.

Sinclair, A., Lee, J., Radley, C., Marzocca, P., Miller, J., Gaviraghi, G., Papazian, A., & Schulze-Makuch, D. (2012). How to develop the solar system and beyond. Leeward Space Foundation.

Smolin, L. (2006). *The trouble with physics*. Penguin Books.

Stein, J. C. (1989). Efficient capital markets, inefficient firms: A model of myopic corporate behavior. *Quarterly Journal of Economics, 104*, 655–669.

Tesla, N. (1900). The problem of increasing human energy. Century Magazine. http://www.teslacollection.com/tesla_articles/1900/century_magazine/nikola_tesla/the_problem_of_increasing_human_energy. Accessed 02 February 2021.

UNFCCC. (2015). Paris agreement. United Nations Framework Convention on Climate Change. https://unfccc.int/sites/default/files/english_paris_agreement.pdf. Accessed 02 December 2020.

UNFCCC. (2021). COP 26 and the Glasgow Financial Alliance for Net Zero (GFANZ). https://racetozero.unfccc.int/wp-content/uploads/2021/04/GFANZ.pdf. Accessed 02 February 2022.

UNPF. (2021). World population dashboard. United Nations Population Fund. https://www.unfpa.org/data/world-population-dashboard. Accessed 02 February 2022.

Urgewald. (2021). Groundbreaking research reveals the financiers of the coal industry. Urgewald. https://urgewald.org/en/medien/groundbreaking-research-reveals-financiers-coal-industry. Accessed 12 December 2021.

Urgewald. (2022). Who is still financing the global coal industry? Urgewald. https://www.coalexit.org/sites/default/files/download_public/GCEL.Finance.Research_urgewald_Media.Briefing_20220209%20%281%29.pdf. Accessed 22 February 2022.

White, F. (1987). *The overview effect—Space exploration and human evolution.* Houghton and Mifflin Co.

5

Quantifying Space Impact

> [O]ne may say that the human ability to understand may be in a certain sense unlimited. But the existing scientific concepts cover always only a very limited part of reality, and the other part that has not yet been understood is infinite. Whenever we proceed from the known into the unknown we may hope to understand, but we may have to learn at the same time a new meaning of the word 'understanding'.
>
> **Werner Heisenberg**, Physics and Philosophy: The Revolution in Modern Science, 1958

> I see Earth! It is so beautiful.
>
> **Yuri Gagarin**, Vostok 1, 1961

Once we have introduced space and its associated stakeholders into the value framework of finance theory and practice, once we have established our responsibility in space and the importance of integrating the space impact of investments into our value framework, the first next step is to quantify the space impact of investments.

As discussed in chapter two and three, whether through externalities or impact investing, quantification of impact has been and still is a very active field of research. The proposed approach and methodology in this chapter complements the literature from a finance perspective, aimed at providing a measurement framework that can integrate the space impact of cash flows and investments into their valuation.

A. V. Papazian, *The Space Value of Money*,
https://doi.org/10.1057/978-1-137-59489-1_5

5.1 The Space Impact Timeline

Let us start by taking a step back and revisiting the most basic element of our discounted cash flow models discussed in chapter two. As shown, the process begins with a timeline that denotes the future expected cash flows that are to be discounted into the present using the relevant discount rate. While this is not a requirement for the application of the equations, it is a visual tool often used when teaching the discounted cash flow models (Brealey et al., 2020).

The first conceptual change we must introduce into our discounted cash flow models and valuation exercises is the double timeline. As in Fig. 5.1, the previous timeline used to measure the discounted risk-adjusted value of future expected cash flows must be expanded, such that, we now have two timelines. One for the risk and time value analysis, where we denote the future expected cash flows, and one for space value analysis, where we denote the investment or cash flow expenditures that are necessary to make those expected cash flows possible.

I call this second timeline, the *Space Timeline*. It is a timeline, because our impact on space also happens over time, and as we track future expected cash flows, we must also track the impact of the investment necessary to create those cash flows.

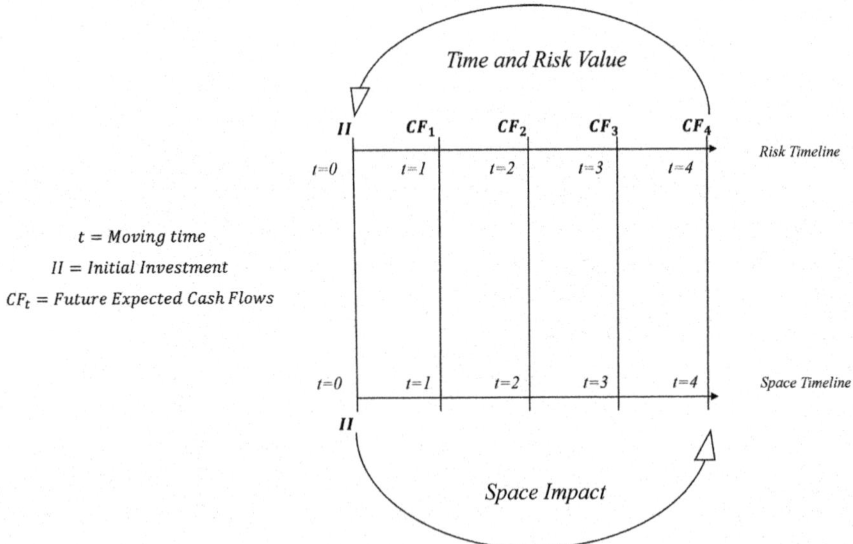

Fig. 5.1 The double timeline: risk and space timelines (*Source* Author)

The purpose of the space timeline is to help visualise the space impact of the investment or cash expenditures that are generating the future expected cash flows. Figure 5.1 is a basic example where the future expected cash flows, CF_t, are generated through the initial investment II.

Our next challenge is to devise the methodology and metrics that can measure the impact of investments and expenditures as we move forward in time, on the space timeline.

Given the discussion in chapter four, and the space value of money principle, we must measure and quantify the space impact of investments in order to ensure we avoid any negative impacts and optimise the positive impacts of investments.

5.2 Space Impact Granularity

Measuring the space impact of an investment involves a closer look at the investment, its origin, and deployment. The below examples shed some light on the many different factors that play a role in defining the space impact of an investment.

The space impact of an investment or cash flow or asset begins when an investment is undertaken as a wire transfer or cash deposit. On a monetary level, the space impact of two investments involving \$1 million dollars are different when one is being financed through a bank loan, which involves new money creation through the banking system, and the other is being financed through a shareholder loan, which does not have the same effect as it involves money already in existence.

The type of the investment or instrument used plays a major role in defining space impact. However, the type of instrument used does not define the impact on its own. For example, an equity investment in a cement factory and an equity investment in a solar energy provider have very different levels and types of space impact. Thus, the instruments used for investment, shares, bonds, loans, convertibles, etc. do not actually define the impact of the investment automatically, it is the activities and production modes used in the value chain of the investment, if any, that define the space impact of an investment. Furthermore, the goods or services involved define the materials used and thus further define the impact.

Moreover, a secondary market equity transaction is different from a primary market transaction, where the former does not involve any cash reaching the company, the latter involves cash being raised directly by the company, affecting its activities, value chain, and space impact. When I purchase shares in Apple Inc. in the secondary market, I am paying another investor who holds the shares, not Apple. Measuring the space impact of an investment must begin by identifying whether the investment is directly supporting a productive value chain or a financial value chain. This is not to suggest that financial value chains are not productive, it is simply to distinguish investments that directly involve the injection of money into a productive activity. A primary market investment in a company, whether in equity or debt form, directly supports the value chain of that company.[1]

In a secondary market stock investment, which does not involve any money reaching the traded company, through the fees paid during the transaction, the investment directly and indirectly supports the value chain of the broker, the bank managing the settlement, and the stock exchange where the share is traded. Thus, a financial value chain can also have a space impact, but a very different one from a direct primary investment in a productive value chain.

Another relevant aspect to consider are the activities involved in and through the investment. Two primary market equity transactions, in two solar energy firms, do not imply identical space impact. Simply, one could be a sales and distribution company selling the technology invented by the other. In other words, the presence of research and development within the value chain of a firm is of great relevance.

Another relevant dimension is the mode of utilisation of specific assets. For example, investing through a primary market equity transaction in two firms producing solar energy technology, both doing R and D work, will still have different space impacts if one is renting a warehouse powered with conventional energy, and the other has built a factory powered by its own solar technology. The created new real estate and its energy consumption will differ from the rented conventional warehouse and its emissions footprint. Although both produce a similar product, a solar energy technology, their space impact will differ.

The type of real assets used and their modes of utilisation and/or creation, as inputs or outputs, are also relevant to measuring the space impact of an investment. For example, gold is an asset but for some investments it is an input, and in others, it is the output. Same goes for oil and gas, and every other intermediate good used in productive activities. Thus, it is important to also distinguish the utilisation of assets and the creation of new assets. These aspects will be explored in detailed in the following sections.

---

[1] In such cases, Porter's (1985) value chain model could be used to track the activities being invested in.

Human capital involvement is another important aspect of the space impact of an investment. For example, let us hypothetically consider the same two primary market equity investments, producing a solar energy technology, both with their own real estate powered by their own technology, but one of them utilises robots to manage the production floor, and the other uses humans. Or, both use humans, but one hires them on part time contracts without pension and the other hires them with full time contracts with pension.

An investment may also involve a governance structure and management process, and when it does, that process may or may not be transparent and fair. An investment may also be layered, exposing the investor to governance factors in multiple entities. For example, investing in a bond issued by company X involves exposure to the management and governance of company X, however, an investment in a bond fund involves exposure to the governance and management of the fund management institution as well as the companies issuing the bonds.

Indeed, the level of granularity is important, and understanding the money, the transaction, and the value chain involved, the assets being created or used, the human capital, and research and development work involved are critical for the process.

Table 5.1 is a sample checklist that summarises the key profile elements of any investment or instrument being analysed for space impact. Naturally, the headings and options listed can be expanded and can be made to include subcategories when and if relevant. For example, depending on the good being produced, material use can be further defined. Depending on real assets used or utilised, as inputs and outputs, supply chain exposures can

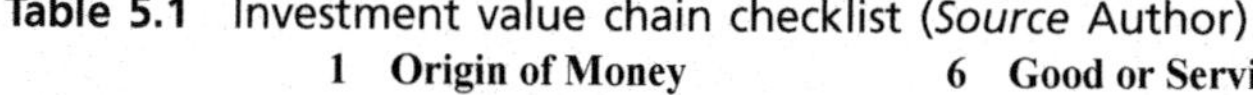

Table 5.1 Investment value chain checklist (*Source* Author)

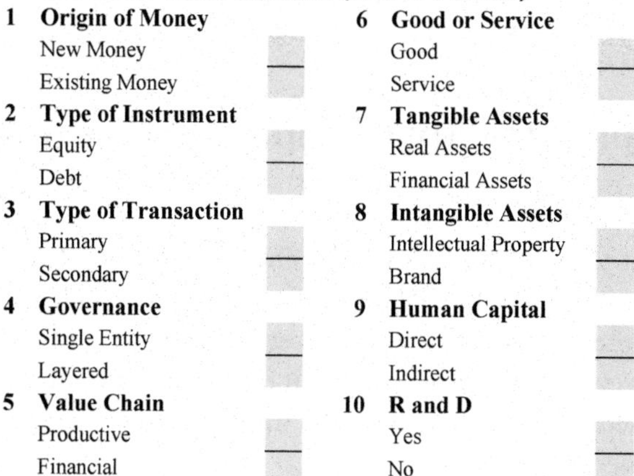

1	**Origin of Money**		6	**Good or Service**	
	New Money			Good	
	Existing Money			Service	
2	**Type of Instrument**		7	**Tangible Assets**	
	Equity			Real Assets	
	Debt			Financial Assets	
3	**Type of Transaction**		8	**Intangible Assets**	
	Primary			Intellectual Property	
	Secondary			Brand	
4	**Governance**		9	**Human Capital**	
	Single Entity			Direct	
	Layered			Indirect	
5	**Value Chain**		10	**R and D**	
	Productive			Yes	
	Financial			No	

be included. The level of detail generated through the initial profiling of an investment opportunity raises the granularity and accuracy of a space impact assessment exercise.

Thus, each investment must first be profiled across 6 main features:

(1) Origin of money: defining whether it is existing money or new money
(2) Instrument type: defining whether it is debt or equity, or other if relevant
(3) Transaction type: defining whether it is a primary or secondary market transaction
(4) Governance type: defining whether it is a single entity or layered
(5) Value chain type: defining whether it is a productive or financial value chain
(6) Type of output: Good or service

Once the above have been identified, and depending on the type of value chain and output, the next step is to identify:

(7) the tangible assets created and/or utilised
(8) the intangible assets created and/or utilised
(9) the human capital involved
(10) the research and development involved

Research and development are categorised separately from assets because under IFRS accounting standards, research, and development are expenses. Thus, it is important to account for them separately. Similarly, human capital, or employment and salaries are treated separately because they are not considered assets in current accounting frameworks.

The next step in the process of space impact evaluation is the mapping of the investment in space.

5.3 Mapping the Space Impact of Investments

When we have profiled the investment according to the previous discussion, the next step is to map the investment and its value chain across the many layers of space it operates in or affects. A company by default is situated on the surface of the planet—through offices or factories and other functional locations. However, depending on the industry and products or services it produces and sells, a company may also have an impact on our hydrosphere or the stratosphere.

For example, two shipping companies A and B, both located in New York, can have a very different impact on different space layers. Besides the continental crust where they are both headquartered, if company A uses ships to move its cargo, and company B uses aeroplanes—one affects the hydrosphere, the other affects the stratosphere. Similarly, a private limousine taxi service will create emissions on the surface of the planet, in the troposphere, a timber company on the other hand, may not be polluting the air as much as causing deforestation, and thus its impact on biodiversity is far more critical than the emissions it generates or causes. Similarly, an aerospace company producing and launching satellites and sending rockets to the moon may also be polluting the upper atmosphere and outer space. A submarine fibre optic cable manufacturer will have to account for its impact on our oceans alongside other aspects.

The first step is to build a heatmap of 'space impact'. Using the same approach briefly introduced in chapter four, this step consists in identifying the layers of space within which companies and/or projects and investments operate or affect. Learning from physics, our understanding of space layers must stretch from the subatomic world of atoms to the cosmic world of planets and galaxies (Greene, 2004; Scott, 2018; Smolin, 2006). For practical purposes, I identify five aspects of space: (1) within: space within matter, (2) inside: inside the planet, (3) surface: the surface of the planet, (4) the planet's atmosphere, and (5) outer space: beyond the planet's atmosphere.[2]

These different aspects of space can be further broken down into layers of space reality within which companies could potentially be operating or having an impact on. Once we have mapped the layers of space within which a company's value chain operates or upon which it has an impact, as both are not necessarily identical, we can understand and conceptualise the many different types of impacts involved. It is only after mapping the space impact of the business and its value chain that we can identify what type of 'negative impacts' we should be looking for, and how we can quantify them.

Figure 5.2 graphically presents the layers of space from within matter to inside the planet, from the surface to the atmosphere, and beyond. Naturally, it can be extended further, the focus here are the layers we have actively engaged with. Table 5.2 describes the same layers of space in a table format, which can be used to map the impact.

While the innermost core, inner core, outer core, and mantle of our planet are not space layers that we have common access to, they must be mentioned and identified when relevant. They are often the subjects of scientific research

[2] Note that space layers are being used to define aspects of space, not dimensions, like length, height, depth, time, and others discussed in physics and string theory (Greene, 2004).

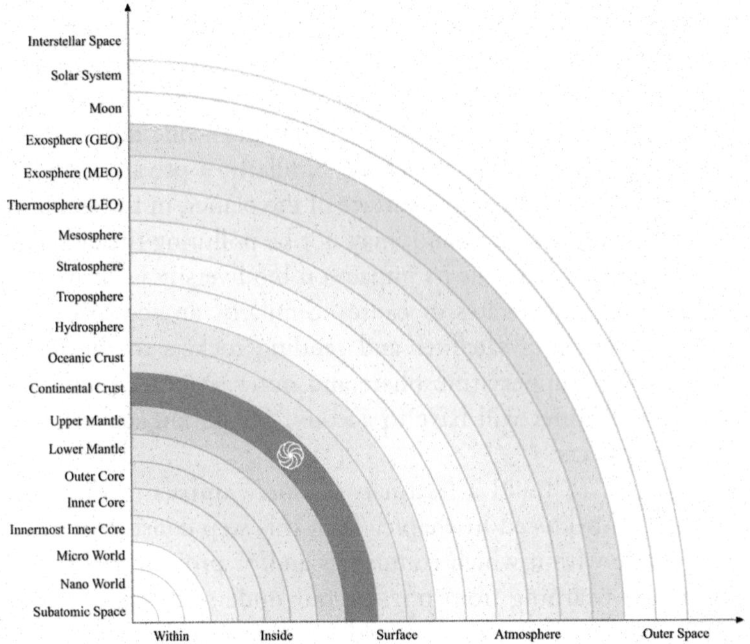

Fig. 5.2 The Space layers from within, inside, surface, atmosphere, to outer space (*Source* Author)

Table 5.2 Space layers (*Source* Author)

Space Continuum	Space Layers	Presence
Outer Space	Interstellar Space	
	Solar System	
	Moon	
Atmosphere	Exosphere: Geostationary orbit (GEO)	
	Exosphere: Medium Earth orbit (MEO)	
	Thermosphere: Low Earth orbit (LEO)	
	Mesosphere	
	Stratosphere	
	Troposphere	
	Hydrosphere	
Surface	Oceanic Crust	
	Continental Crust	֍
	Upper Mantle	
	Lower Mantle	
Inside	Outer Core	
	Inner Core	
	Innermost Inner Core	
	Micro World	
Within	Nano World	
	Subatomic Space	

through seismic waves and are thus important to identify. Naturally, in very few cases we expect to see human activity affecting those specific layers.

The symbol ❁ in Table 5.2 and the grey area in Fig. 5.2 signify the space layer where all companies are located, the continental crust, the land surface where all investments are initiated and implemented. All projects, companies, cash flows, and assets are present and affecting this layer. Other layers require an analysis and mapping of space impact by looking at the value chain and activities of the investment in question.

The layered analysis of space impact is important from four main perspectives. On the most basic level, it is about mapping the value chains of investments according to the layers of space they affect. While all companies are located on land, the continental crust, not all make use of our oceans through shipping or fishing, and impact our hydrosphere. While all have the continental crust as a base of operations, not all deliver their service in the lower stratosphere transporting people and cargo across continents.

The next level of relevance of a layered approach to space impact is that the intensity and nature of impact differs across layers. For example, GHG emissions in the stratosphere and emissions on the surface of the planet have very different implications. Indeed, the effects of height on emissions behaviour has been observed and studied from a variety of perspectives (Muralikrishna & Manickam, 2017; Vallero, 2019; Van der Hoven, 1975).

The layered approach to space impact analysis is also important because cleaning the same pollutant in different space layers involves different technologies. Cleaning plastic from the oceans and cleaning plastic from our streets are two very different processes, involving different technologies. Cleaning waste in orbit around the planet is a very different technological challenge than cleaning waste from our rivers.

Last but not least, given all of the above, the costs of impact differ across different layers of space, even for the same pollutant or type of waste.

Naturally, each one of the identified space layers can be further divided into sublayers. In Table 5.3, two commonly used and abused layers of space are presented in more detail. The continental crust and our hydrosphere are most commonly used for all sorts of human activity. From agriculture to mining to fishing and transport, we have numerous industrial and productive activities that make use of the land and water on our planet. Table 5.3 expands on oceanic zones and soil layers, as well as provides examples of lakes and vegetation that can have specific features that must be taken into account in different investments and value chains. Chart 5.1 is a sample hypothetical distribution histogram of space impact. I will explore the many aspects

Table 5.3 Space layers further details: Hydrosphere and continental crust (*Source* Author)

Space Layers	Sub-Layers	Sub-Layer Type Examples
	Seas, Lakes, Rivers, Ice Sheets	Tectonic lakes
		Volcanic lakes
	Oceans	Glacial lakes
	Epipelagic Zone - The Sunlight Zone	Fluvial lakes
	Mesopelagic Zone - The Twilight Zone	Solution lakes
Hydrosphere	Bathypelagic Zone - The Midnight Zone	Landslide lakes
	Abyssopelagic Zone - The Abyss	Aeolian lakes
	Hadal Zone - The Trenches	Shoreline lakes
		Organic lakes
		Anthropogenic lakes
		Meteorite lakes
	Land Surface	Tundra
	Mountains	Taiga
	Built Up	Temperate broadleaf and mixed forest
		Temperate steppe
		Subtropical moist forest
	Vegetation	Mediterranean vegetation
Continental Crust		Tropical and subtropical moist forests
	Cropland	Arid desert
	Soil	Xeric shrubland
	O Horizon - Organic Layer	Dry steppe
	A Horizon - Top Soil Nutrient Layer	Semiarid desert
	E Horizon - Eluviation Layer	Grass savanna
	B Horizon - Subsoil Mineral Layer	Tree savanna
	C Horizon - Regolith Layer	Tropical and subtropical dry forest
	R Horizon - Bedrock Layer	Tropical rainforest
	Deep Crust	Alpine tundra
		Montane forest

of space impact and the equations through which we can conceptualise and calculate said impact in the following sections.

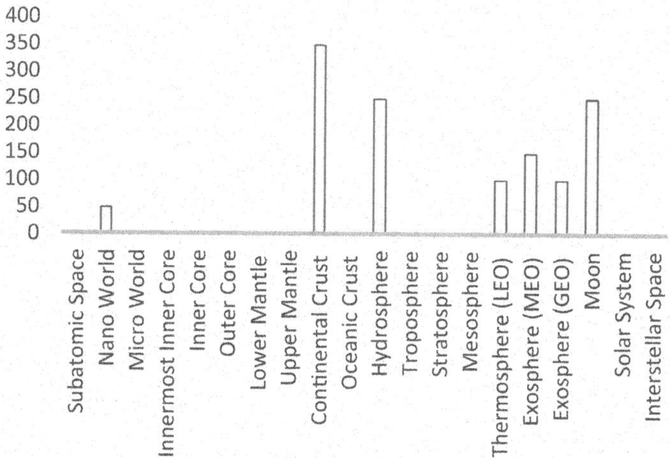

Chart 5.1 Sample Space Impact Distribution Histogram by Space Layer for a hypothetical Aerospace Investment that uses nanotechnology, launches satellites into orbit, lands rockets in the ocean, and has a moon habitat project in process (*Source* Author)

5.4 Quantifying Space Impact

Once we have mapped the space impact of an investment and its value chain, we must quantify it across all the layers of space within which the investment operates or affects. This must be done before we can integrate that impact into the monetary value attached to the investment, or business. The impact of an investment on space takes many shapes and forms. Indeed, investments vary in the level and type of impact they have on space. The most important point to clarify here is that when looking for the space impact of an investment or business we should not be looking for its negative impact alone. Investments may generate pollution and waste, but they may also create value. Thus, an authentic and objective space impact assessment must include all types of impact. This is necessary because, as we discussed in chapter two, space impact is omitted from our value models, thus, it is not just pollution and waste that matter, but the economic impact of cash flows as well.

Looking at the space impact of investments must be conceptualised and explored from the perspective of the two new stakeholders of finance, the planet and humanity. Starting off with the pollution-averse planet, we must identify and establish the pollution and waste that the investment may be causing, whether CO_2 or Methane, or plastic, or radioactive waste. Furthermore, from the planet's perspective, the investment's impact on biodiversity is of crucial importance.

From the perspective of the eternal and aspirational human society, several aspects come into play, namely, the human capital impact, the R and D impact, the new asset impact and the new money impact. Moreover, all of these aspects are affected by the governance factors that influence the investment and the value chains attached to the investment. To start off at the highest conceptual level, the aggregate space impact of an investment, which I have previously called Gross/Net Space Value (Papazian, 2011, 2017, 2021) can be summarised as the planetary, human and economic impact of the investment across all space layers and time periods considered.

$$NSV_{T\&S} = \{Planetary,\ Human,\ and\ Economic\ Impact\}_{All\ Space\ Layers\ \&\ Time\ Periods}$$

$$NSV_{T\&S} = Net\ Space\ Value\ of\ Investment$$

$$T = Total\ Number\ of\ Years\ of\ Investment\ being\ Considered$$

$$S = All\ Space\ layers\ Involved\ in\ the\ Investment$$

$$\begin{aligned} NSV_{T\&S} = {} & \sum_{t=1}^{T}\sum_{s=1}^{S} Pollution\ \&\ Biodiversity\ Impact \\ & + \sum_{t=1}^{T}\sum_{s=1}^{S} Human\ Capital\ \&\ R\ and\ D\ Impact \\ & + \sum_{t=1}^{T}\sum_{s=1}^{S} New\ Asset\ \&\ New\ Money\ Impact \end{aligned} \tag{5.1}$$

This is a conceptual summary of the equation, describing the net space value of an investment, or the aggregate space impact, as the sum of its planetary, human, and economic impacts across the time and space horizon of the investment. These three aspects of impact are in their turn divided into the Pollution, Biodiversity, Human Capital, R and D, New Asset and New Money impacts of the investment across the time horizon and space layers involved. Each of these aspects is discussed in more detailed in the following sections.

5.4.1 Planetary Impact: Pollution, Waste, Nature and Biodiversity

5.4.1.1 Pollution and Waste

Pollution is a multifaceted reality. It can occur through human action and natural phenomena, like a volcanic eruption spreading volcanic ash over thousands of miles. In this discussion, we are concerned with anthropogenic pollution, i.e., pollution that has human activity as a source. Pollution is caused by a variety of processes that include industrial and production activities, as well as consumption and transportation activities. The encyclopaedia Britannica defines pollution as follows:

> [T]he addition of any substance (solid, liquid, or gas) or any form of energy (such as heat, sound, or radioactivity) to the environment at a rate faster than it can be dispersed, diluted, decomposed, recycled, or stored in some harmless form. The major kinds of pollution, usually classified by environment, are air pollution, water pollution, and land pollution. Modern society is also concerned about specific types of pollutants, such as noise pollution, light pollution, and plastic pollution. Pollution of all kinds can have negative effects on the environment and wildlife and often impacts human health and well-being (Britannica, 2021).

In this section, we will consider pollutants to be those materials, substances, elements, particles, nano-particles, and molecules that are known to harm the environment and human and animal well-being and are generated through the design, testing, industrial or non-industrial production, transportation, distribution, consumption, and disposal of final and intermediary products and services across all sectors of the economy. As such, waste pollution is included in the discussion.

We must clearly state here that the pollutants must be identified, considered and reported whether they are officially listed and identified by relevant regulatory agencies or not. The 'forever chemicals' scandal that revealed gross misconduct by one of the world's most renowned chemical manufacturers is an example of what not to do. Unfortunately, despite laws and corporate social responsibility rhetoric, we have corporations that dispose toxic chemicals in leaking landfills near human and animal habitats and hide such activities from those affected and the authorities (Benesh, 2020; Lerner, 2015; Sisk, 2020).

This is why the space value of money principle and the associated metrics are key to the transformation we seek, and not just in terms of emissions and chemicals, but all types of pollutants and waste materials, including plastic in the ocean, debris in outer space, or radioactive waste hidden in the ocean floor or landfills.

When considering the pollution impact of investments, naturally, the literature on negative environmental externalities can be very useful and relevant. Furthermore, the Life Cycle Assessment approach to plot and identify such impacts for products and services remains valid (EIA, 1995; Nguyen et al.,

2016; SETAC, 1991, 1993). The metric provided here is an aggregation tool for the quantification of pollution impact.

The pollution impact of an investment across all space layers (S) and over the lifetime of the investment (T) can be quantified by multiplying the pollutants that are being created multiplied by the cost of removal or clean-up. Naturally, the cost of clean-up must be per pollutant, per space layer. This is so because the cost of clean-up for a pollutant may vary across space layers and at different times. These costs can be calculated either directly, or through proxies like recycling costs (Trucost, 2013; WRAP, 2018). In the case of Carbon, the clean-up cost should not be confused with Carbon Prices (Ellerman et al., 2010) or the Effective Carbon Rate, as defined by the OECD (OECD, 2018, 2019), although such mechanisms could be used as proxies when relevant. For GHG emissions and conversion rates, the UK government's conversion factors could be used to standardise GHG quantities and costs (HM Government, 2019, 2021).

$$Pollution\ Impact_{All\ Pollutants\ Across\ Time\ and\ Space} = PI_{T,S,P} =$$

$$PI_{T,S,P} = Sum\ of\ All\ Individual\ Pollutant\ Impacts\ Across\ Years\ and\ Space\ Layers$$

$$PI_{T,S,P} = \sum_{t=1}^{T} \sum_{s=1}^{S} \sum_{p=1}^{P} Q_{pst} \times C_{pst} \tag{5.2}$$

$T = Total\ Number\ of\ Years\ in\ the\ Investment$

$P = Number/Types\ of\ Pollutants\ Involved$

$S = All\ Space\ Layers\ Involved$

$Q_{pst} = Quantity\ of\ Pollutant\ (p)\ in\ time\ (t)\ in\ space\ layer\ (s)$

$C_{pst} = Cost\ of\ Cleanup\ for\ Pollutant\ (p)\ in\ time\ (t)\ in\ space\ layer\ (s)$

The impact of a specific pollutant (Individual Pollutant Impact—IPI) must be calculated taking into account the fact that the quantity and cost of clean-up may differ across the space layers and time periods involved in the investment. For example, cleaning a specific pollutant from land surfaces may be cheaper than cleaning the same pollutant from the ocean floor.

$$IPI_{T,S,P_1} = \sum_{t=1}^{T} \sum_{s=1}^{S} Q_{p_1 st} \times C_{p_1 st} \tag{5.3}$$

$IPI_{T,S,P} = Individual\ Pollutant\ Impact\ Across\ Years\ and\ Space\ Layers$

If the investment is absorbing the pollutant, so it is cleaning it up, the measure still applies, but not as a loss of value, but as a gain. Which implies that the positive or negative sign attached to this measure depends on whether the investment is creating pollution or cleaning up pollution. In the case where a pollution causes permanent damage without any measure of clean-up, an exaggerated cost could be used to act as a prevention mechanism. This does not actually reverse the damage that has already been caused but can prevent future projects with such an effect to ever be undertaken.

Given that a company could be creating or causing many different types of pollution, in many different layers of space, or even within one space layer, to quantify the total space impact we need to apply this to all levels of activity and pollutants across all the layers of space within which the company is operating. If a company is operating within *one layer of space*, and let's assume that it is creating *one pollutant* over the time period of the investment T, and the cost of clean-up, C, is *constant across the years*, then its overall Pollution Impact (PI) is equal to the Individual Pollutant Impact in the layer of space identified and the cost of clean-up in that layer:

$$PI_{T,S,P} = IPI_{T,S_1,P_1} = C_{p_1 s_1 t_1} \times \sum_{t=1}^{T} Q_{p_1 s_1 t} \tag{5.4}$$

If there is *one pollutant*, but it is affecting three different layers of space, with three different clean-up costs, over a T period.

$$PI_{T,S,P} = IPI_{T,S_{1,2,3},P_1} = \sum_{t=1}^{T} \sum_{s=1}^{3} Q_{p_1 st} \times C_{p_1 st} \tag{5.5}$$

Note that in the above equation, p_1 is a specific pollutant, while s identifies the three involved space layers $s_{1,2,3}$ across all the years t of the investment.

If there are many different pollutants being created in a specific layer of space, and we are interested in isolating the pollution impact of the investment on a specific layer of space, then we can use:

$$PI_{T,S_1,P} = \sum_{t=1}^{T} \sum_{p=1}^{P} Q_{ps_1 t} \times C_{ps_1 t} \tag{5.6}$$

$$S_1 = \textit{The Space Layer of Concern}$$
$$P = \textit{Number/Types of Pollutants Involved in Space Layer } (s_1)$$
$$Q_{ps_1t} = \textit{Quantity of Pollutant } (p) \textit{ in Period } (t) \textit{ in space layer } (s_1)$$
$$C_{ps_1t} = \textit{Cost of Clean - up of Pollutant } (p) \textit{ in Period } (t) \textit{ in Space Layer } (s_1)$$

Note that in the above equation, s_1 is a specific layer, and no more a variable index, while p and t are indices identifying a specific year and a specific pollutant or waste.

5.4.1.2 Biodiversity

While pollution and waste are critical for our assessment of planetary impact, they are not the only aspects that must be considered. The biodiversity impact is of crucial importance (IPBES, 2019). Our agricultural and consumer good production processes have resulted in expanding land use that has caused extensive biodiversity loss across the planet. From food to timber production, as well as infrastructure developments, biodiversity loss has grown hand in hand with economic expansion, affecting countries and regions differently.

White et al. (2021) summarise the biodiversity challenge and the many reports and studies that provide the relevant evidence eloquently, I believe a direct quote is more appropriate here. They write:

> Despite increasing recognition of its importance, biodiversity is in precipitous decline (Díaz et al., 2019; Tittensor et al., 2014). Recent reports estimate that 75% of the terrestrial environment and 66% of the marine environment have been severely altered by human activity (Halpern et al., 2015; IPBES, 2019; Venter et al., 2016), and that between 1970 and 2014 populations of monitored species have declined by an average of 70% (WWF, 2018b). This decline is largely driven by the continued growth of the global economy (Hooke et al., 2012; IPBES, 2019; Maxwell et al., 2016). From aquaculture and forestry to mining, consumer goods, and infrastructure, industrial development across sectors is closely tied to biodiversity loss. Business operations and supply chains act to increase the production and movement of goods, often at the expense of natural ecosystems through increasing habitat loss, fragmentation, pollution, invasive species introductions, and overexploitation (Díaz et al., 2019; Krausmann et al., 2017). Consequently, biodiversity loss is recognized as a major global challenge for the private sector presenting operational, financial, and reputational risks (Global Canopy & Vivid Economics, 2020; WEF, 2021). (White et al., 2021)

The Ecological Society of America (ESA, 1997) identifies three types of biodiversity, genetic, species, and ecosystem biodiversity, and six types of threats: (1) habitat loss and destruction, (2) altered composition, (3) introduction of exotic species, (4) overexploitation, (5) pollution and contamination, (6) climate change.

Whatever the type of biodiversity loss we face through an investment, whatever the specific causes of loss, whatever the specific threats involved, whatever the specific environments they affect, i.e., marine, land, air, and even one day other planets, and whatever the level of diversity within those habitats lost, for the purposes of an all-encompassing conceptual metric for biodiversity loss measurement in monetary terms, the concept of biodiversity impact must be measured at the most abstract level, i.e., the area of space affected.

Unlike the logic applied to pollution and waste impact quantification, we cannot allocate a cost of clean-up for biodiversity loss. However, we can allocate a cost of restoration or habitat replacement costs (Environment Agency, 2015) as an indicative measure of the biodiversity cost of the investment. While there is an extensive literature on the feasibility of habitat restoration (HM Government, 1994; Parker, 1995), a similar concept of Habitat Restoration Cost (HRC) has been discussed at length in the natural capital literature (Pearce & Moran, 1994; Shepherd et al., 1999).

Naturally, restoration does not imply the ability to bring back a species that has gone extinct. However, biodiversity loss is the driver of species extinction, and thus, any impact on any area or environment must be accounted for. Newmark et al. (2017) observe the time lag involved between loss and extinction and identify the opportunity for restoration and rehabilitation that can avoid extinction. When an investment can potentially cause the extinction of a species, it must be prevented from doing so, and thus, the restoration cost of habitats must account for such impacts on animals, insects, and plants. This is why the density and diversity of specific areas are relevant variables when measuring restoration costs.

In a 1999 English Nature report, Shepherd et al. (1999) dig deeper into the subject and offer their approach which could be used to estimate the projected cost per hectare. Similarly, a recent report by the European Commission Joint Research Centre Institute for Environment and Sustainability (Dietzel & Maes, 2015) provides cost estimates of specific restoration measures.

Diversity has been recognised to be a driving factor of biodiversity loss assessment. From hotspots (Mittermeier et al., 1998; Myers, 1989) to the STAR system and heat indices (Hawthorne, 1996; Hawthorne & Abu-Juam,

1995), which aim to classify different species and genetic pools affected within a specific site or area, are all important in the quantification of the cost of restoration and the assessment of extinction risk caused by any specific investment or supply chain. This is why the importance of the site or space being affected comes into play (CISL, 2020, discussed in chapter three). In another paper titled 'Integrating global and local values: A review of biodiversity assessment', published by the International Institute for Environment and Development (IIED), Vermeulen and Koziell (2002) provide an important overview of key issues to consider when assessing the restoration costs for a specific habitat.

While the restoration costs are themselves a subject of extensive research, in the equation below I summarise the concept at a level of abstraction that must be applied given specific investment characteristics and the value chains involved. Given the space value of money principle, the purpose here is to prevent negative biodiversity impact by integrating their projected restoration costs in the value of expected future cash flows. Thus, a summary abstract measure is used.

$$Biodiveristy\ Impact_{Across\ Time\ and\ Space} = BI_{T,S,B} =$$

$$BI_{T,S,B} = \sum_{t=1}^{T}\sum_{s=1}^{S}\sum_{b=1}^{B} A_{bst} \times R_{bst} \tag{5.7}$$

$T =$ *Total Number of Years in the Investment*

$S =$ *All Space Layers Involved*

$B =$ *All Types of Biodiversity Involved*

$BI_{T,S} =$ *Sum of All Biodiversity Impacts Across Years and Space Layers*

$A_{bst} = Area(ha)$ *of Biodiversity Impact* (b) *in time* (t) *in space layer* (s)

$R_{bst} =$ *Cost of Restoration* $\left(\frac{\$}{ha}\right)$ *of Biodiveristy Impact* (b) *in* (t) *in* (s)

If the investment is restoring a habitat and biodiversity, the measure still applies, not as a loss of value, but as a gain. Which implies that the positive or negative sign attached to this measure depends on whether the investment is causing or restoring biodiversity loss. In the case where an investment causes permanent irreversible damage without any measure of restoration, an exaggerated cost could be used to act as a prevention mechanism. This does not actually reverse the damage that has already been caused but can prevent future projects with such an effect to ever be undertaken.

Biodiversity can affect land surfaces, but also our hydrosphere, as such, specific attention must be paid to ensure that the relevant surfaces or areas are identified across different layers of space and are measured in the same unit as the cost of restoration. In some cases and situations, a volume based measure may be more appropriate.

Naturally, as it has been well-documented and argued before, prevention of biodiversity loss is the right approach and far more cost effective than restoration. This methodology does not advocate for restoration, it simply provides a methodology to quantify the biodiversity loss involved in an investment to identify negative impacts, for their eventual integration in our value models.

For example, if an investment involves one type of biodiversity impact within two different layers of space over a period of time T, we can aggregate the biodiversity impact as:

$$Biodiveristy\ Impact_{Across\ Time\ and\ Space\ Layers} = BI_{T,S_{1,2},B_1} =$$

$$BI_{T,S_{1,2},B_1} = \sum_{t=1}^{T}\sum_{s=1}^{2} A_{b_1 st} \times R_{b_1 st} \tag{5.8}$$

$T = Total\ Number\ of\ Years\ in\ the\ Investment$

$S_{1,2} = All\ Space\ Layers\ Involved$

$B = All\ Types\ of\ Biodiversity\ Involved$

$BI_{T,S,B_1} = Sum\ of\ All\ Biodiversity\ Impacts\ b_1\ Across\ Years\ and\ Space\ Layers$

$A_{st} = Land\ Area\ (ha)\ of\ Biodiversity\ Impact\ in\ time\ (t)\ in\ space\ layer\ (s)$

$R_{st} = Cost\ of\ Restoration\left(\frac{\$}{ha}\right)\ of\ Biodiveristy\ Impact\ in\ (t)\ in\ (s)$

To conclude the planetary impact section, we must put the two main equations together and provide an aggregate for the planetary impact of the investment. The equation is given below:

$$Planetary\ Impact = PLANETI = Pollution\ Impact + Biodiveristy\ Impact$$

$$PLANETI_{T,S} = PI_{T,S,P} + BI_{T,S,B}$$

$$PLANETI_{T,S} = \sum_{t=1}^{T}\sum_{s=1}^{S}\sum_{p=1}^{P} Q_{pst} \times C_{pst} + \sum_{t=1}^{T}\sum_{s=1}^{S}\sum_{b=1}^{B} A_{bst} \times R_{bst} \tag{5.9}$$

5.4.2 Human Impact: Human Capital and R and D

5.4.2.1 Human Capital

Human capital, the cornerstone of all productive processes on our planet, is a fundamental aspect of any investment. While some investments are more or less human capital intensive, using technology, hardware or software, to different levels, human capital remains an essential aspect of all investments done by humans.

This section discusses human capital impact outside of pollution, waste, and biodiversity. The negative environmental impacts of investments, which affect human well-being as well, have been discussed already in the previous section. Here, I focus on quantifying the impact on humans, on society, conceptualised as human capital impact.

The World Bank through the Human Capital Project launched the Human Capital Index in 2018. Focused on country level data and analysis, the premise of the index and the project is the following:

> Human capital consists of the knowledge, skills, and health that people accumulate over their lives. People's health and education have undeniable intrinsic value, and human capital also enables people to realize their potential as productive members of society. More human capital is associated with higher earnings for people, higher income for countries, and stronger cohesion in societies. It is a central driver of sustainable growth and poverty reduction. (World Bank, 2020, 1)

Human capital is involved in almost all investment transactions, however minimally or marginally. Even an entirely technology based secondary market transaction in the stock market involves human capital, even if not visible to the trader at the moment of purchase or investment. Even when an investment, like a secondary market transaction in the stock market, does not reach the company as working capital or cash flows, humans are still involved and affected, directly and indirectly. This is so because the transaction, even if minimally, supports the value chains of the broker, the banker, and the stock exchange.

Similarly, when we save or invest in a blocked savings account, we may not be investing in a productive value chain, but we are taking part or indirectly affecting the value chain of the account providing savings institution, which in turn could be investing or lending to other productive value chains involving human capital. Thus, the first step in identifying the human capital impact of an investment is to identify the nature of the connection.

A direct connection involves an investment that is providing cash flows to a productive value chain where human capital is employed, and an indirect connection is where the cash flows or fees involved may affect human capital after investment, due to the follow up use of the fees/investment.

Quantifying indirect human capital impact is extremely challenging because one cannot know the possible uses of the saved or invested money, or how parallel investments would be executed based on a blocked savings account that provides necessary liquidity to a bank. Quantifying direct human capital impact is relatively simpler, as this would involve scrutinising the value chain that is being invested in.

Not all investments involve human capital involvement or deployment, and when they do, they do not all have the same type of deployment and use. Some are fair, others not, some include wider support mechanisms, education and health expenditures, others not. Still, some invite the involvement of foreign talent and workers, thus enriching the local workforce, others not. Moreover, some invest in the wider community they are connected to, others not, some compensate their employees for work related damages or costs, others not.

In our current accounting standards, employment or the human capital used by a company, is not categorised as an asset. Employment is an expenditure, combining salaries, benefits, national insurance payments, pension payments, etc. which together reveal the human capital impact of an investment. Thus, the human capital impact of an investment can be derived from its employment expenditure, using payroll, as well as all spending on human capital, whether in the form of benefits, or facilities and support.

I have summarised such expenditures by grouping them into five subcategories, ETHICS, representing Employment, Training, Health, Immigration, Compensation, and Social Investment Expenditures. Although the vast majority of humans live and work on the continental crust of our planet, some are on ships and submarines in our hydrosphere, and since the year 2000, a small group of exceptional pioneers have been living in the International Space Station (ISS) in low Earth orbit. Thus, a multi-layered space analysis is relevant. I expect that over the next many decades, we will be able to extend and expand our reach, and this analysis will include many more space layers, such as the moon and eventually Mars.

$$Human\ Capital\ Impact = HCI_{T,S}$$

$$HCI_{T,S} = Fair\ and\ responsbile\ expenditure\ on\ humans\ within\ and\ around$$

$$HCI_{T,S} = f \times \sum_{t=1}^{T} \sum_{s=1}^{S} E_{st} + T_{st} + H_{st} + I_{st} + C_{st} + S_{st} \tag{5.10}$$

$$\begin{aligned}
S &= \textit{All Space Layers Involved} \\
T &= \textit{Total Number of Years in the Investment} \\
f &= \textit{Coefficient of Fairness} \\
E_{st} &= \textit{Employment Expendiuture} \\
T_{st} &= \textit{Training and Education Expendiuture} \\
H_{st} &= \textit{Health Related Expendiuture} \\
I_{st} &= \textit{Immigration Related Expendiuture} \\
C_{st} &= \textit{Compensation Expenditure} \\
S_{st} &= \textit{Social Investment Expendiuture}
\end{aligned}$$

The coefficient f is a number between $-1 \leq f \geq 1$, that identifies the level of fairness in the organisation and/or in the employment of the human capital being used in the investment. This can be defined by gender balance, equal pay, fair treatment of employees, sick pay, proper pension management, etc. The value assigned to f in a specific investment does require the further analysis of the operations and processes that govern the employment and deployment of human capital through the investment.

The coefficient f introduces the opportunity to qualify employment and related expenditures by a value that reflects the fairness in the human capital management of the investment in question. A fair and well-managed human capital expenditure that respects our requirements for gender balance, fair wages, equal pay, and fair treatment can receive a value of 1. An investment where the human capital process is neither fair nor equal can receive a value of zero, or a value less than 1. In some cases, where the human capital process is exploitative, causing mental health issues, it can be given a value of -1.

5.4.2.2 Research and Development

The next type of human impact to consider is research and development. All our innovative technologies are born out of an intensive research and development process. Ideas and insights materialise into economic value through the efforts put into researching and developing them into actual products and services. Indeed, this is true even when the subject matter is pure philosophy or metaphysical in nature. In other words, R and D is central to any process of transformation, and critical for human evolution and growth.

Given the intensive process of transformation implied in the transition to a low carbon economy, and given the ongoing digital transformation, we are moving into the most R and D intensive phase of the world economy. From carbon capture technologies to new artificial intelligence solutions and the digitisation of entire value chains, R and D investments are and will continue to grow in importance.

In a recent OECD report, the Frascati Manual, R and D expenditure is defined as:

> Research and experimental development (R&D) comprise creative and systematic work undertaken in order to increase the stock of knowledge—including knowledge of humankind, culture and society—and to devise new applications of available knowledge.... The term R&D covers three types of activity: basic research, applied research and experimental development. (OECD, 2015, 44).

OECD (2015) identifies R and D projects as the conceptual unit that helps quantify a specific set of R and D activity or activities. Whatever the goal or field involved in a specific R and D project (See Table 5.4), the main idea here is to understand that an investment can involve one, two, or more R and D projects, which may or may not be in the same field and may or may not be focused on the same objective. We adopt this approach here to facilitate the quantification of R and D impact.

When considering the R and D expenditure of projects, we must take into account the current and fixed capital expenditures (OECD, 2015, 30). When considering labour costs, we must take into account the fact that employment expenditures are also treated as part of the human capital expenditure discussed in the previous section.

$$R\&D\ Impact = RDI_{T,S,N}$$

$$RDI_{T,S,N} = \sum_{t=1}^{T}\sum_{s=1}^{S}\sum_{n=1}^{N} h_n \times RD_{tsn} \tag{5.11}$$

$h_n = Coefficient\ of\ health\ of\ R\ and\ D\ Project\ (n)$

$RD_{T,S,N} = R\&D\ Expenditure\ per\ Project\ N\ across\ T\ and\ space\ layers\ S$

$S = All\ Space\ Layers\ Involved$

$T = Total\ Number\ of\ Years\ in\ the\ Investment$

$N = Number/All\ R\ and\ D\ Projects\ Involved\ in\ the\ Investment$

Table 5.4 OECD fields of R and D classification (OECD, 2015)

Broad classification	Second-level classification
1. Natural sciences	1.1 Mathematics 1.2 Computer and information sciences 1.3 Physical sciences 1.4 Chemical sciences 1.5 Earth and related environmental sciences 1.6 Biological sciences 1.7 Other natural sciences
2. Engineering and technology	2.1 Civil engineering 2.2 Electrical engineering, electronic engineering, information engineering 2.3 Mechanical engineering 2.4 Chemical engineering 2.5 Materials engineering 2.6 Medical engineering 2.7 Environmental engineering 2.8 Environmental biotechnology 2.9 Industrial biotechnology 2.10 Nano-technology 2.11 Other engineering and technologies
3. Medical and health sciences	3.1 Basic medicine 3.2 Clinical medicine 3.3 Health sciences 3.4 Medical biotechnology 3.5 Other medical science
4. Agricultural and veterinary sciences	4.1 Agriculture, forestry, and fisheries 4.2 Animal and dairy science 4.3 Veterinary science 4.4 Agricultural biotechnology 4.5 Other agricultural sciences
5. Social sciences	5.1 Psychology and cognitive sciences 5.2 Economics and business 5.3 Education 5.4 Sociology 5.5 Law 5.6 Political science 5.7 Social and economic geography 5.8 Media and communications 5.9 Other social sciences
6. Humanities and the arts	6.1 History and archaeology 6.2 Languages and literature 6.3 Philosophy, ethics and religion 6.4 Arts (arts, history of arts, performing arts, music) 6.5 Other humanities

R and D impact can be negative. It is very much possible to witness an R and D process that damages the environment and human health. The extensive evidence and legal cases around Teflon and C8, or 'eternal chemicals' in general, demonstrate that R and D can be a healthy or unhealthy process. Besides the environmental and biodiversity damages caused by toxic landfills, the R and D process could be unhealthy for those involved and their families, depending on industry and practice. Thus, a coefficient of health must be included when assessing the impact of R and D projects and expenditures.

The coefficient h_n is included in the equation as a number equal to either -1 or 1, to identify the healthy management of the R and D project being assessed, taking into account safety and protection measures for the environment, the individuals involved, and their families. This coefficient is either positive or negative without any degrees in between given that the process is either healthy or not.

R and D activity creates value even when it does not lead to a final product and/or service. Indeed, based on the methodology of space impact assessment described in this chapter, the finished products and technologies are accounted for in the asset section. Here, R and D expenditure is taken into account as a value creating activity that enhances knowledge, experience, and stretches the boundaries of knowledge and/or a productive or business process.

Bringing the two elements of human impact together, Human Capital Impact and R and D Impact, we can conceptualise the total as follows:

$$Human\ Impact = HUMANI = Human\ Capital\ Impact + R\&D\ Impact$$

$$HUMANI = HCI_{T,S} + RDI_{T,S,N}$$

$$HUMANI = f \times \sum_{t=1}^{T}\sum_{s=1}^{S} E_{st} + T_{st} + H_{st} + I_{st} + C_{st} + S_{st} + \sum_{t=1}^{T}\sum_{s=1}^{S}\sum_{n=1}^{N} h_n \times RD_{tsn} \tag{5.12}$$

5.4.3 Economic Impact: New Assets and New Money

When we discussed the omission of space and our responsibility for space impact in chapter two, it was clear and evident that conventional finance models and equations have also ignored to take into account the economic impact of cash flows and investments.

While the discussion to this point has focused on planetary and human impact, we should also introduce the assessment of space impact from an economic standpoint, by looking at the new assets and new money that an investment creates. After all, investments may have the same level of emissions, low or high, the same level of risk and time adjusted returns, but very different asset and money impacts.

5.4.3.1 New Assets

For example, consider two projects, both requiring $1 million dollars of investment, having the same payback period, the same risk, the same return on investment, and both spending $300,000 on real estate—do they have the same space impact? Not really, because the first investment is using the $300,000 to rent its real estate in the city over the lifetime of the investment, and the second investment is buying a small plot of land outside the city and developing its own solar energy powered real estate. Similarly, two companies that are in the solar energy business, one is developing its own new technology, the other is a distribution outlet. Both happen to have the same return on investment, but they have very different asset impacts given the creation of new intellectual property and new technology.

One important clarification is due. While an assessment of the asset impact is crucial, asset creation *does not absolve* an investment of the responsibility for any negative planetary or human impact. This is a fundamental point to address and integrate into the analysis. The space value of money principle requires that our investments avoid negative impacts and optimise positive space impacts. Offsetting positive and negative impacts is not an acceptable 'solution'.

The next step in the quantification of space impact is the identification of all the new assets that the investment will create across all the layers of space. The logic here is identical to the previous sections, however, the elements being considered are different. We quantify the assets created across the timeline of the investment as well as the space layers affected.

While identifying the assets created through the investment may be a relatively simple exercise for new investments, for existing companies and projects, the task is more complex given that many assets could be only partly financed by the new investment. However, assuming that an investment opportunity is being assessed based on a clear investment proposal, we could identify and apply the relevant proportions.

To calculate the new assets created through an investment, we can apply the following formula, where we add the value of all tangible and intangible assets created across all space layers:

$$New\ Asset\ Impact = NAI_{D,S,A} = \sum_{s=1}^{S}\sum_{a=1}^{A} k_a \times BVA_{asD} \qquad (5.13)$$

$k_a = $ *Coefficient of Transition Value*

$S = $ *All Space Layers Involved*

$D = $ *Period/Date When Asset is Created and Added in Books*

$A = $ *All Tangible and Intangible Asset Created through the Investment*

$BVA_{asD} = $ *Book Value of Asset* (a) *in space layer* (s) *recorded at date* (D)

This exercise must be done for all types of assets, tangible and intangible, real and financial, to extract a detailed view of the asset impact of the investment, giving us the total value of assets created through the investment throughout the years, across all space layers. In some specific cases, we can use the market value of assets, instead of their book value, depending on the type and volatility of the created asset. Note that the assets are added once, as per the date (D) when they are added on the books.

In order to contextualise the importance of this aspect, consider two aerospace companies A and B. A is focused on taking tourists to low orbit, and B is focused on mining asteroids following the US Senate legislation HR.2262, passed in November 2015, which states that 'any asteroid resources obtained in outer space are the property of the entity that obtained them, which shall be entitled to all property rights to them, consistent with applicable federal law and existing international obligations' (US Senate, 2015). Whatever the relative investment amounts and other types of impacts involved, the potential for new asset creation in aerospace company B is immensely greater than the one in A.[3]

K_a, the coefficient of transition is defined on the individual asset level, and it is a coefficient with a value between $-1 \leq k_a \geq 1$ and it can be used to qualify the assets vis-a-vis the transition. For example, consider the hypothetical case of two companies C and D, where C is planning to build a new coal mine and/or a new oil refinery, and D is creating a new carbon capture technology and plant. The assets created by these two firms cannot be considered at face value. Given the transition, the assets created by company C will soon become obsolete and they may even become a liability over time. This is the concept of 'stranded assets' defined as 'assets [that] suffer from unanticipated or premature write-offs, downward revaluations or are converted to liabilities' (Caldecott et al., 2013, 7). To reflect this fact, we could assign k_{Ca} a value of zero, and k_{Da} a value of one. Note that *k* is defined at an individual asset level not company level, and this is just an example.

Another aspect to consider when quantifying assets is the nature of the assets. Consider two manufacturing firms, E and F, where E is manufacturing mobile ventilators for hospitals and care homes, and company F is manufacturing arms and bombs. Company E's inventory saves lives, company F's inventory destroys lives and can even destroy other assets and damage the environment. We cannot possibly value these inventories the same way. A similar analysis can be applied to two companies producing solar panels and barrels of oil, their inventories have very different space impacts. Thus, the coefficient of transition allows the investor to integrate that judgement into the new asset impact of the investment in question.[4]

---

[3] For a fascinating account of the mining possibilities and challenges outside of our planet see Lewis (2011, 1996) and Goswami and Garretson (2020).

[4] Naturally, the value of the coefficient is assigned according to the investor's assessment of conditions and timelines. In some situations, manufacturing defence technology can be considered a positive value added and in the current stage of the transition, existing fossil fuel investments cannot be immediately discounted to zero until alternative infrastructures and technologies have been put in place. Thus, the value of the coefficient of transition should be a matter of degrees given conditions and timelines.

Alongside creating assets, investments can or may also cause the creation of new money in the economy. Just like the asset, biodiversity, human capital, R and D, pollution, and waste impacts cannot be ignored, the monetary impact of an investment is also part of its space impact and should be given proper consideration.

5.4.3.2 New Money

Just like the asset impact of an investment, the monetary impact of an investment, even when significant, does not absolve the company or investment from the responsibility of its negative impact. In other words, the monetary impact and the asset impact cannot and do not cancel out the negative impact of an investment. This is crucial to understand and establish.

Quantifying the monetary impact of an investment begins at the source. In some cases, an investment is achieved through a bank loan or other forms of debt or leverage, and the money being invested is itself new money. Thus, its new money impact is equal to the investment amount to start with.

When assessing the monetary impact of an investment, the first step is to identify whether the invested money is new money or already existing money. An equity investment from a private equity firm does not involve money creation, it involves the transfer of money from one account to another. A shareholder loan that goes into supporting the working capital of a firm does not involve new money creation. A bond purchase by a private investor supporting the debt raise of a corporation in the primary market does not involve new money creation. On the other hand, a bank loan (through central or commercial banks) does involve the creation of new money, along with other cases of leverage where the loan is creating the money through the banking system.[5]

Thus, to quantify the monetary impact of an investment, we must first clarify if the investment is new money in the form of bank debt.[6] The equation below defines the proportion of the investment that is new money by multiplying the initial investment with the debt ratio and the bank loan ratio. This is so because the investment could be partly financed through other means besides bank debt.

$$New\ Money\ Impact = NMI_T = II \times DR \times BLR \tag{5.14}$$

[5] I discuss central bank and commercial bank new money creation in detail in chapter eight.

[6] We could also consider the scenario where the initial investment is a foreign investment implying direct capital inflow from the start, thus the entire investment would be new money.

$$II = Initial\ Investment,\ DR = Debt\ Ratio,\ BLR = Bank\ Loan\ Ratio$$

$$T = Total\ across\ Years\ in\ the\ Investment$$

If we were to treat the initial investment as a new deposit, whether partly or entirely new money, or entirely existing money, we could also measure its impact on the money creation cycle as it is spent within the economy, becoming new deposits in other banks. To measure how much money creation potential the investment implies as it is spent in the economy, if it is, we can multiply it with the *actual* money multiplier.[7] This is proposed knowing full well the limitations of the money multiplier concept. In this equation, the money multiplier is the actual multiplier as observed in the economy at time t-1, and it is used as a descriptive measure.

If we wanted to gauge the potential for money creation as the investment is spent in the economy, if it is, we must also account for planned imports and expected exports through the investment as those will take away or add to the initial investment's monetary impact in the local economy. Imports will involve part of the initial investment being sent abroad, while exports will involve the reverse inflow into the macro-economy through foreign exchange.

Thus, we could also expand the equation to include the money creation potential of the investment as it is spent, by adding the expected exports, subtracting the planned imports, and then multiplying the sum $(II + X - M)$ by the money multiplier. Given that the monetary impact, as of today, happens only in one space layer, and given that it has to be considered across the entire investment time window, it is written as follows:

$$New\ Money\ Impact = NMI_T$$

$$NMI_T = (II \times DR \times BLR) + mm \times (II + X_T - M_T) \tag{5.15}$$

$mm = Money\ Multiplier\ DR = Debt\ Ratio\ BLR = Bank\ Loan\ Ratio$

$M_T = Planned\ Imports\ X_T = Expected\ Exports\ II = Initial\ Investment$

$N = Total\ Years\ in\ the\ Investment$

---

[7] We use the actual money multiplier as observed in t-1, as a descriptive measure. Recent studies have argued about the irrelevance of the money multiplier for central bank policy transmission, also clarifying the relationship between the money multiplier and bank credit policy. Thus, this is only used as an after the fact observation, not a policy tool (Ihrig et al., 2021; McLeay et al., 2014).

The new money being created through an investment is an important measure to consider when looking at large investments. Moreover, it is a relevant dimension when considering the exports and imports involved through an investment and/or when designing and deploying public investments. Once again, we must clarify that a high positive new money impact does not absolve the investment of its responsibilities for negative space impact.

To put together the economic impact elements in one integrated equation:

$$ECONOMIC\ Impact = ECONOMICI = New\ Money\ Impact + New\ Asset\ Impact$$

$$ECONOMICI = NMI_T + NAI_{D,S,A}$$

$$ECONOMICI = (II \times DR \times BLR) + mm \times (II + X_T - M_T) + \sum_{s=1}^{S}\sum_{a=1}^{A} k_a \times BVA_{asD} \tag{5.16}$$

5.4.4 Governance

Governance is a key thematic focus of the recently growing discussion in sustainable finance, and ESG ratings and factor integration have played an important role in operationalising the role of governance in the field. However, the subject of good governance has been of relevance for both governments as well as private and public corporations, for many decades. Governance has been considered a critical variable in understanding the issue of government failure and corruption, and a key dimension of markets through direct links to competition. Governance has also acquired more importance since the publication of the Sustainable Development Goals. As Goal 16 states, accountable, inclusive, and effective institutions at all levels is a key to the just world the UN SDGs aim to build.

> Promote peaceful and inclusive societies for sustainable development, provide access to justice for all and build effective, accountable and inclusive institutions at all levels. (UNDESA, 2021)

Meanwhile, corporate governance has been a central topic in finance and management research for some time. Managerial incentives, management turnover, board composition, succession, committees, contracts, shareholding structures, and a host of other aspects have been studied at length, all ultimately looking at good governance and agency costs associated with corporations and the pursuit of shareholder value (Jensen & Meckling, 1976; Fama, 1980; Fama & Jensen, 1983; Shleifer & Vishny, 1989, 2012; Gilson, 1989; Gordon & Becker, 1964; Bluedorn, 1982; Brown, 1982; Brickley & Drunen, 1990; Dissanaike & Papazian, 2005; Bhagat et al., 2008; Daines et al., 2010).

Without going into an in-depth literature review of corporate governance and its many direct and indirect indicators, and their correlations with performance, the main point being made here is that governance plays a key role in the way an investment is managed, which affects both the *outcomes* and *impacts* of the investment. In other words, governance affects both, the expected future cash flows, and the space impact of the investment.

Volkswagen's emissions test scandal is a prime example (Murphy, 2019). Thus, from an impact assessment perspective, the key purpose of the governance factor is to reveal or assess the reliability and veracity of the facts and figures being reported—facts and figures upon which the entire analysis of impact is based.

The proposed methodology here is to assess or qualify the results based on a governance coefficient that reflects the level of transparency and accountability in the investment, whether involving a primary or secondary market transaction, whether involving an investment with an attached financial or productive value chain.

I propose the coefficient g to denote the investor's degree of belief in the truthfulness of the reported figures and facts, and the degree of trust in the authenticity of commitment to targets and objectives.

$$g = \textit{Coefficient of Good Governance}$$

The coefficient g may or may not be considered relevant to a specific investment. When it is, it should be a coefficient between $0 \leq g \leq 1$ reflecting the level of trust in the veracity of the figures based on the corporate governance measures and structures in the investment. Transparency and accountability could be measured or quantified through the coefficient g through a number of different metrics. The Valuing Respect project and the resources and metrics published by Shift (2021) are one such useful tool that could be used when assessing the culture of governance involved in an investment.

Naturally, the value of g ascribed to an investment or company may vary based on the elements considered and whether or not the investment in question concerns an existing business with a past record, or a new one that has no proven governance practices in place yet. For new investments, and to give the investment the benefit of the doubt, g can be equal to 1, if clear principles are defined and declared through the investment proposition. In their absence, a value of less than 1 could be considered. When assessing existing businesses, declared principles and institutionalised procedures should be qualified by an assessment of actual track records.

When quantifying the governance coefficient, we must also consider the possibility of a layered exposure. When an investment is made into an equity fund that invests in global equities, the investor is exposed to both the governance process of the asset manager who is managing the fund, but also the governance process of the companies whose stocks are being invested in through the fund. In a direct primary equity investment in a company, investors are exposed to the governance structure of the business they are investing in. However, if they invest through funds and/or asset management firms, or through a syndicated investment programme managed by a broker or investment bank, they are exposed to the governance processes of more than one entity. When the investment is exposed to two different governance processes or structures, each could be qualified with a different g_1 and g_2 coefficient.

Once governance has been identified as a factor, the next step is to identify whether the exposure is to a single entity or multiple entities. In both cases, the main purpose of the coefficient g is to establish the degree of trust in the facts and figures and plans involved in the investment, and it is defined through the accountability and transparency of the governance process.

5.5 Net Space Value or Aggregating Space Impact

We started this discussion by drawing the space timeline alongside the risk timeline to help measure the space impact of an investment. We then identified the need to build the basic profile of the investment, to identify the source of money, the transaction type, the value chain involved, the assets being created or utilised, the human capital, and R and D expenditures. Subsequently, we revealed the need to identify the space impact map of the investment in order to understand and identify the space layers affected by the investment. Finally, we identified three aspects of space impact, planetary, human, and economic, and proposed an approach for their quantification.

The aggregate space impact of the investment is the sum of all the previously discussed impacts, covering Pollution, Biodiversity, Human Capital, Research and Development, New Assets, and New Money. This is calculated across all the layers of space where the investment operates or affects.

Furthermore, we identified the governance factor g, which may or may not be relevant, as a coefficient of trust in the facts, figures, plans and targets being assessed by the investor.

When aggregating space impact, we must remember that all the elements of Net Space Value can be either positive or negative, depending on the activities involved. In other words, if an investment is capturing emissions or restoring a habitat, its PI and BI impacts would be positive. Otherwise, they would be negative. In some specific cases, other categories of impact may also result in a negative number. This measure must be calculated or projected across the lifetime of the investment before actual investment. For opportunities that already exist, the impact can be calculated looking back three or five years, depending on history and availability.

$$Net\ Space\ Value = Aggregate\ Space\ Impact$$

$$Net\ Space\ Value_{T,S} = g \times \left(PI_{T,S,P} + BI_{T,S,B} + HCI_{T,S} + RDI_{T,S,N} + NAI_{D,S,A} + NMI_{T}\right) \quad (5.17)$$

$$g = Coefficient\ of\ Good\ Governance$$
$$PI_{T,S,P} = Pollution\ and\ Waste\ Impact$$
$$BI_{T,S,B} = Biodiveristy\ Impact$$
$$HCI_{T,S} = Human\ Capital\ Impact$$
$$RDI_{T,S,N} = Research\ and\ Development\ Impact$$
$$NAI_{D,S,A} = New\ Asset\ Impact$$
$$NMI_{T} = New\ Money\ Impact$$
$$S = All\ Space\ Layers\ Involved$$
$$T = Total\ Number\ of\ Years\ in\ the\ Investment$$

$$Net\ Space\ Value + II = Gross\ Space\ Value$$

$$NSV + II = GSV \quad (5.18)$$

When all space impacts are equal to zero, i.e., when Net Space Value is zero, then the Gross Space Value equation would give us:

$$GrossSpaceValue_{T,S} = 0 + II \quad (5.19)$$

$$GSV = II$$

Which is, indeed, the equality that is the very bottom threshold stated in the principle. A dollar invested in space must, at the very least, have a dollar's worth of positive impact on space. When an investment is at the minimum threshold, its Net Space Value is equal to zero, and its Gross Space Value is equal to the initial investment.

5.6 Space Impact Types and Intensity

As discussed previously and hypothetically described in Fig. 5.3, the space impact of investments could be divided into three parts: (1) planetary impact, accounting for pollution and biodiversity impact; (2) human impact, accounting for human capital and research and development impact, and (3) economic impact, accounting for new asset and new money impact. Investments can have a diverse footprint when it comes to these three types of impact (Table 5.5), as such, expected future cash flows can have three types of impact intensities. The introduction of these measures fine tunes the concept and analysis of impact and reveals the possibility of a more granular assessment.

Through the space impact calculations and the above metrics, we now recognise that every expected cash flow in the future involves a space impact that makes it possible. Thus, we can also calculate the space impact of an investment at different time intervals and identify the impact intensity of the expected cash flows.

$$Impact\ Intensity_t = Space\ Impact\ Itensity\ of\ Expected\ Cash\ Flow_t$$

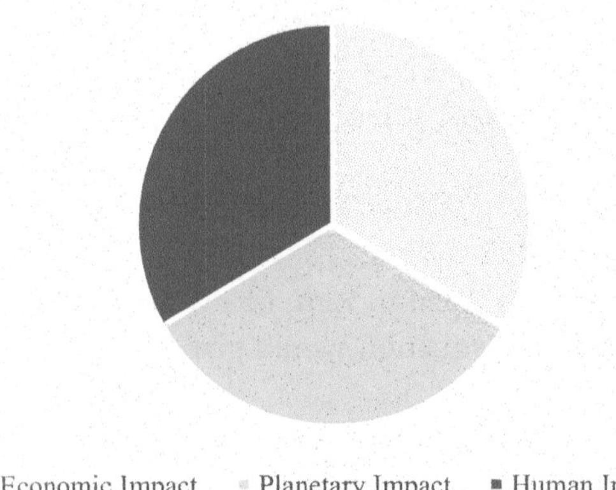

Fig. 5.3 The three aspects of space impact (*Source* Author)

Table 5.5 Net space value components and equations

Impact Aspect	Net Space Value	$g \times (PI_{T,S,P} + BI_{T,S,B} + HCI_{T,S} + RDI_{T,S,N} + NAI_{D,S,A} + NMI_T)$
PLANETARY	Pollution Impact	$PI_{T,S,P} = \sum_{t=1}^{T}\sum_{s=1}^{S}\sum_{p=1}^{P} Q_{pst} \times C_{pst}$
	Biodiversity Impact	$BI_{T,S,B} = \sum_{t=1}^{T}\sum_{s=1}^{S}\sum_{b=1}^{B} A_{bst} \times R_{bst}$
HUMAN	Human Capital Impact	$HCI_{T,S} = f \times \sum_{t=1}^{T}\sum_{s=1}^{S} E_{st} + T_{st} + H_{st} + I_{st} + C_{st} + S_{st}$
	R and D Impact	$RDI_{T,S,N} = \sum_{t=1}^{T}\sum_{s=1}^{S}\sum_{n=1}^{N} h_n \times RD_{tsn}$
ECONOMIC	New Asset Impact	$NAI_{D,S,A} = \sum_{s=1}^{S}\sum_{a=1}^{A} k_a \times BVA_{asD}$
	New Money Impact	$NMI_T = (II \times DR \times BLR) + mm \times (II + X_T - M_T)$
Coefficients	Fairness	f
	Health	h
	Transition	k
	Governance	g

Table 5.6 Planetary, economic, human impact intensity measures (*Source* Author)

	Planetary impact	Economic impact	Human impact
Equation	$PI_{t,S,P} + BI_{t,S,B}$	$NAI_{d,S,A} + NMI_t$	$HCI_{t,S} + RDI_{t,S,N}$
Signs	+, 0, -	+, 0, -	+, 0, -
Positive	Cleans pollutions and waste and restores biodiversity	Creates new assets and new money	Allocates resources to a healthy and effective human capital and/or R and D process
Negative	Adds pollution and waste or causes biodiversity loss	Destroys existing assets and/or existing money	An unhealthy and toxic R and D process or Human Capital Management
Null	It neither pollutes nor cleans, neither destroys nor restores	It neither creates new assets or new money, nor destroys existing assets or money	It does not allocate resources to human capital and/or R and D
Impact Intensity of Expected CF	$\frac{PI_{t,S,P} + BI_{t,S,B}}{ExpectedCashFlow_t}$	$\frac{NAI_{d,S,A} + NMI_t}{ExpectedCashFlow_t}$	$\frac{HCI_{t,S} + RDI_{t,S,N}}{ExpectedCashFlow_t}$
High Ratio	A high planetary impact intensity implies that the investment involves high pollution/waste or biodiversity impact per future expected cash flows, negative or positive	A high economic impact intensity implies that the investment involves high asset and/or money impact per future expected cash flow, negative or positive	A high human impact intensity implies that the investment involves high human capital or R and D expenditure per future expected cash flow, negative or positive
Low Ratio	A low planetary impact intensity implies that the expected cash flows have a very low pollution/waste and or biodiversity impact, positive or negative	A low economic impact intensity implies that the expected cash flows involve low asset or money impact, positive or negative	A low human impact intensity implies that the expected cash flows involve low human capital or research development impact, positive or negative

	Planetary impact	Economic impact	Human impact
Sign	The direction of impact is defined by the signs of the numerator	The direction of impact is defined by the signs of the numerator	The direction of impact is defined by the signs of the numerator
	Example Scenarios		
	High Negative	**Low Positive**	**Low Positive**
Example Case 1	An investment in an existing coal mine will have no new asset or money creation impact, low human capital and R and D expenditure, and yet, will have a high emissions impact leading to a higher negative planetary impact intensity		
	Low Positive	**Low Positive**	**Low Positive**
Example Case 2	An investment in a solar power distribution company through a secondary market transaction. Low on pollution and waste, low on new asset creation, and low on human and R and D expenditure		
	High Positive	**High Positive**	**High Positive**
Example Case 3	A primary market investment in a solar power technology development company that runs on solar power and invests in human capital and R and D for its product development and maintenance		
	High Negative	**High Positive**	**High Negative**
Example Case 4	An investment in a chemical manufacturer that runs on solar power with low emissions has new plants and equipment built and developed, but has a toxic R and D process and dumps chemicals damaging staff, biodiversity, and human life around its toxic landfills		

$$\text{Impact Intensity}_t = \frac{\text{Net Space Value}_t}{\text{Expected Cash Flow}_t} \tag{5.20}$$

This measure is not a decision metric, it is a descriptive measure that reveals the amount of impact involved in generating the cash flows promised by the investment. It differs from the Carbon Intensity measure described in chapter three. A low carbon intensity investment can imply both a high or low space impact intensity in other aspects across the three main categories of impact discussed earlier, i.e., planetary, human, and economic.

The Impact Intensity of an investment and the Carbon Intensity of an investment may or may not be correlated. This nuance is important to consider and understand. Carbon intensity measures emissions per sales, or in this case, emissions vis-à-vis expected cash flows.

$$\text{Carbon Intensity}_t = \frac{\text{Emissions}_t}{\text{Expected Cash Flow}_t} \tag{5.21}$$

Given the methodology of space impact discussed in this chapter, a very green investment with zero emissions could have a very large space impact through massive new asset creation. Indeed, to achieve our transition to a low carbon economy, we must have more investments that have a low carbon intensity but very high positive impact intensity.

This is important and not often discussed in the current debate on ESG factors or portfolio alignment using temperature rise methods. In the race to Net Zero, we must be able to distinguish between two investments that have the same level of carbon intensity or implied temperature score but have very different impact intensities.

$$\text{Planetary Impact Intensity}_t = \text{PI and BI per Expected Cash Flow}_t$$

$$\text{PLANETII}_t = \frac{PI_{t,S,P} + BI_{t,S,B}}{\text{Expected Cash Flow}_t} \tag{5.22}$$

$$\text{Human Impact Intensity}_t = \text{HCI and RDI per Expected Cash Flow}_t$$

$$\text{HUMANII}_t = \frac{HCI_{t,S} + RDI_{t,S,N}}{\text{Expected Cash Flow}_t} \tag{5.23}$$

$$\text{Economic Impact Intensity}_t = \text{NAI and NMI per Expected Cash Flow}_t$$

$$ECONOMICII_t = \frac{NAI_{d,S,A} + NMI_t}{ExpectedCashFlow_t} \tag{5.24}$$

When calculating the economic impact intensity of the cash flows, we consider only those new assets created during the relevant year ($NAI_{d,S,A}$), and consider the yearly new money impact on a pro rata basis (NMI_t).

While it is far less obvious and harder to identify investments that destroy existing assets or remove existing money, they do exist, and the economic impact of an investment can also be positive, negative, or null. For example, when an investment involves the early repayment of an existing bank loan, or when an investment involves the purchase and breakdown of an asset to be sold as parts. In such cases, while the cash flow return to the individual investor may be positive, the economic asset impact could be negative. Similarly, the pollution, biodiversity, human capital, and R and D impacts can be negative when they cause pollution and biodiversity loss, and when they are toxic and unfair causing health and well-being concerns.

Table 5.6 summarises these measures and provides simple examples and scenarios. I have not included the coefficient g in these measures to focus on the key propositions—g could be added when governance is critical and relevant. Moreover, it must be noted here that these are not decision rules on their own, they do not actually provide any decision-making logic yet, they do however provide a detailed quantitative description of the space impact of an investment. As I will discuss in the next chapter, these metrics are critical for any impact analysis that aims to integrate impact into our value equations.

Indeed, after the space impact assessment exercise has been completed and the three impacts quantified, the next step is to visualise the planetary, human, and economic impact intensity and the signs of impact. This basic matrix is an easy and convenient way to identify the transition type involved in a specific investment.

As we move to Net Zero, which is mainly an emissions target, our purpose is to reduce negative impact by reducing emissions. Within the space value of money framework, negative impacts are not acceptable, and must be phased out and reduced to zero, and all low positive impacts must be optimised to high, if we are to achieve the transition we have set as a target, if we are to truly transition to a healthier planet and economy.

The space value of money principle sets a high positive impact intensity target for all categories of impact. Thus, in Table 5.7, the target for all investments is set to be a high positive impact across all categories. However, as the principle states, the responsibility of space impact is to also avoid negative

Table 5.7 IMIX—Impact Intensity Matrix (*Source* Author)

		Planetary Impact	Human Impact	Economic Impact	Transition
Positive	*High*	֍	֍	֍	Target
	Low				Transition 3
0					
	Low				Transition 2
Negative	*High*				Transition 1

Table 5.8 Example IMIX of an actual investment (*Source* Author)

		Planetary Impact	Human Impact	Economic Impact	Transition
Positive	*High*			X	Target
	Low		X		Transition 3
0					
	Low				Transition 2
Negative	*High*	X			Transition 1

impact, and thus, the acceptable threshold of impact is NSV = 0, or GSV = II.

Naturally, the actual impact intensity and sign of impact for different investments may not be on target and may vary across different types of impact (See Table 5.8). A high negative planetary impact investment may have a high positive economic impact, and a low positive human impact. Indeed, until recently and given the value framework applied in finance, such investments have been tolerated given their investment returns.

5.7 The Space Growth Rate

Beyond the specific intensities of space impact involved in an investment, as an aggregate measure, space impact denotes both positive and negative impacts. While pollution is unwanted, asset creation is a definite value added for the wider economy. While biodiversity loss is to be avoided and negative, healthy R and D expenditure is desirable and a key force of innovation and evolution in our society. Such nuances of impact analysis and measurement improve our understanding and assessment.

However, it is also important to summarise the analysis of space impact on a broader level where all these impacts are aggregated. This is achieved through the space growth rate. When we have calculated the aggregate space impact of an investment across its lifetime, and we know the investment amount at stake, we can calculate the space growth rate of the initial investment through the below formula. The space growth rate (SPR) is the implied annual rate that takes us from the Initial Investment (II) in the present, to the aggregate space impact, or Net Space Value (NSV) across the T periods of the investment:

$$SPR = \sqrt[T]{\frac{NSV_{T,S}}{II}} - 1 \tag{5.25}$$

$$\begin{aligned}
&SPR = \textit{The Space Growth Rate per period} \\
&II = \textit{Initial Investment} \\
&T = \textit{Total Number of Years in the Investment} \\
&NSV_{T,S} = \textit{Net Space Value} \\
&\textit{Net Space Value}_{T,S} = g \times \big(PI_{T,S,P} + BI_{T,S,B} + HCI_{T,S} \\
&\qquad\qquad + RDI_{T,S,N} + NAI_{D,S,A} + NMI_{T}\big)
\end{aligned}$$

The growth rate, in its current form, does not discriminate against the types of impact. However, if we were to break it down, we could also reveal the space growth attached to each type of impact. Note that the space growth rate is relevant as a benchmark measure when the Pollution Impact (PI) and Biodiversity Impact (BI), New Asset (NAI) and New Money Impact (NMI), and Human Capital (HCI), and R and D Impact (RDI) are positive (Table 5.9).

5.8 Conclusion

For many decades, finance theory and industry have been focused one measuring the risk and time value of cash flows and assets. The analysis of balance sheets, income statements, and cash flow statements were all geared towards confirming future expectations of performance in terms of profitability, not impact on space, on the planet, on society, and our climate.

This chapter introduced the steps and equations through which we can profile, map, and quantify the space impact of investments across all space layers involved in an investment. Starting from an expanded timeline to include space impact analysis, the chapter introduced the concepts of planetary, human, and economic impact, and their corresponding intensity measures. Each of these aspects was further explored through their component elements, i.e., Pollution, Biodiversity, Human Capital, R and D, New Asset, and New Money impacts. Building on these metrics, I also introduced the concepts of Net Space Value and Gross Space Value, and the

Table 5.9 Planetary, economic, human impact space growth rates (*Source* Author)

	Planetary Impact Growth Rate - SPR_P	Economic Impact Growth Rate - SPR_E	Human Impact Growth Rate - SPR_H
Impact Equations	$PI_{T,S,P} + BI_{T,S,B}$	$NAI_{D,S,A} + NMI_T$	$HCI_{T,S} + RDI_{T,S,N}$
Signs	+ , 0, −	+ , 0, −	+ , 0, −
Annual Rate Equations	$\sqrt[T]{\frac{PI_{T,S,P}+BI_{T,S,B}}{II}} - 1$	$\sqrt[T]{\frac{NAI_{D,S,A}+NMI_T}{II}} - 1$	$\sqrt[T]{\frac{HCI_{T,S}+RDI_{T,S,N}}{II}} - 1$
Positive Rate	Adding value by extracting pollution or restoring biodiversity	Adding value by creating new assets and/or new money	Adding value by spending on human capital and R and D expenditure
Negative Rate	Unacceptable—An investment with a negative pollution/waste or biodiversity impact per future expected cash flow is not acceptable under SVM	Unacceptable—An investment with a negative asset and money impact per future expected cash flow is not acceptable under SVM	Unacceptable—An investment with a negative Human Capital and R and D impact per future expected cash flow is not acceptable under SVM
Optimisation Target	≥ 0	≥ 0	≥ 0

Space Growth Rate. The discussion in this chapter was entirely focused on the metrics through which we can quantify space impact.

The next chapter will integrate the space impact of investments into our value equations and propose a decision-making logic that will allow us to, at the very least, prevent the inception of new investments and companies that will require transition. This is done by ensuring that investments with negative space impact are identified and either improved before investment or disqualified from investment as they should be.

References

Benesh, M. (2020). Why are DuPont and Chemours still discharging the most notorious 'forever chemical'? Environmental Working Group. https://www.ewg.org/news-insights/news/why-are-dupont-and-chemours-still-discharging-most-notorious-forever-chemical. Accessed 12 December 2021.

Bhagat, S., Bolton, B., & Romano, R. (2008). The promise and Peril of corporate governance indices. *Colombia Law Review*, *108*(8), 1903–1882. https://www.jstor.org/stable/40041812. Accessed 15 February 2022.

Bluedorn, A. (1982). Theories of turnover: Causes, effects, and meanings. In S. B. Bacharach (Ed.), *Research in the sociology of organisations* (pp. 75–128). JAI Press.

Brickley, J., & Drunen, L. (1990). Internal corporate restructuring: An empirical analysis. *Journal of Accounting and Economics, 12*, 251–280. https://doi.org/10.1016/0165-4101(90)90050-E. Accessed 02 February 2021.

Britannica. (2021). Environmental pollution. Encyclopaedia Britannica. https://www.britannica.com/science/pollution-environment. Accessed 02 February 2022.

Brealey, A. R., Myers, C. S., & Allen, F. (2020). *Principles of corporate finance* (13th ed.). McGraw Hill.

Brown, M. C. (1982). Administrative succession and organisational performance: The succession effect. *Administrative Science Quarterly, 27*, 1–16. https://doi.org/10.2307/2392543. Accessed 02 February 2021.

Caldecott, B., Howarth, N., & McSharry, P. (2013). Stranded assets in agriculture: Protecting value from environment-related risks. Smith School of Enterprise and the Environment. https://www.smithschool.ox.ac.uk/publications/reports/stranded-assets-agriculture-report-final.pdf. Accessed 02 February 2021.

CISL. (2020). Measuring business impacts on nature: A framework to support better stewardship of biodiversity in global supply chains. Cambridge Institute for Sustainability Leadership, Cambridge. https://www.cisl.cam.ac.uk/system/files/documents/measuring-business-impacts-on-nature.pdf. Accessed 02 February 2022.

Daines, R. M., Gow, I. D., & Larker, D. F. (2010). Rating the ratings: How good are commercial governance ratings? *Journal of Financial Economics, 98*(3), 439–461. https://doi.org/10.1016/j.jfineco.2010.06.005. Accessed 02 February 2021.

Díaz, S., et al. (2019). Pervasive human-driven decline of life on Earth points to the need for transformative change. *Science, 366*(6471), eaax3100. https://doi.org/10.1126/science.aax3100. Accessed 02 February 2021.

Dietzel, A., & Maes, J. (2015). Costs of restoration measures in the EU based on an assessment of LIFE projects. European Commission. https://publications.jrc.ec.europa.eu/repository/bitstream/JRC97635/lb-na-27494-en-n.pdf. Accessed 12 January 2020.

Dissanaike, G., & Papazian, A. V. (2005). Management turnover in stock market winners and losers: A clinical investigation. European Corporate Governance Institute. ECGI. Finance Working Paper N° 61/2004. https://papers.ssrn.com/sol3/papers.cfm?abstract_id=628382. Accessed 02 February 2021.

EA. (2015). Cost estimation for habitat creation—Summary of evidence. Environment Agency. https://assets.publishing.service.gov.uk/media/6034ef5ee90e0766033f2ea7/Cost_estimation_for_habitat_creation.pdf. Accessed 02 June 2020.

EIA. (1995). Electricity generation and environmental externalities: Case studies. Energy Information Administration. https://www.nrc.gov/docs/ML1402/ML14029A023.pdf. Accessed 02 February 2022.

ESA. (2017). Biodiveristy. Ecological Society of America. https://www.esa.org/wp-content/uploads/2012/12/biodiversity.pdf. Accessed 02 February 2022.

Fama, E. F. (1980). Agency problems and the theory of the firm. *Journal of Political Economy*, 660–672. https://www.jstor.org/stable/1837292. Accessed 02 February 2021.

Fama, E. R., & Jensen, M. C. (1983). Separation of ownership and control. *Journal of Law and Economics*, 301–325. https://www.jstor.org/stable/725104. Accessed 02 February 2021.

Gilson, S. (1989). Management turnover and financial distress. *Journal of Financial Economics, 25*, 241–262. https://doi.org/10.1016/0304-405X(89)90083-4. Accessed 02 February 2021.

Global Canopy, Vivid Economics. (2020). The case for a task force on nature-related financial disclosures. https://globalcanopy.org/wp-content/uploads/2020/11/Task-Force-on-Nature-related-Financial-Disclosures-Full-Report.pdf. Accessed 12 December 2021.

Goswami, N., & Garretson, P. (2020). *Scramble for the skies: The great power competition to control the resources of outer space*. Lexington Books.

Greene, B. (2004). *The fabric of the Cosmos: Space, time, and the texture of reality*. Penguin Books.

Halpern, B. S., et al. (2015). Spatial and temporal changes in cumulative human impacts on the world's ocean. *Nature Communications, 6*(1), 1–7. https://doi.org/10.1038/ncomms8615

Hawthorne, W. D., & Abu-Juam, M. (1995). *Forest protection in Ghana*. IUCN.

Hawthorne, W. D. (1996). Holes and the sums of parts in Ghanaian forest: Regeneration scale and sustainable use. *Proceedings of the Royal Society of Edinburgh, 10B*, 75–176.

Heisenberg, W. (1958). *Physics and philosophy: The revolution in modern science*. Penguin Books.

HM Government. (2019). 2019 Government greenhouse gas conversion factors for company reporting: Methodology paper for emission factors final report. UK Government. Department for Business, Energy, and Industrial Policy. https://assets.publishing.service.gov.uk/government/uploads/system/uploads/attachment_data/file/904215/2019-ghg-conversion-factors-methodology-v01-02.pdf. Accessed 02 February 2020.

HM Government. (2021). Greenhouse gas reporting: Conversion factors 2021. UK Government. https://www.gov.uk/government/publications/greenhouse-gas-reporting-conversion-factors-2021. Accessed 02 February 2022.

HM Government. (1994). Biodiversity: The UK action plan. Her Majesty's Stationery Office (HMSO), Office of Public Sector Information. https://data.jncc.gov.uk/data/cb0ef1c9-2325-4d17-9f87-a5c84fe400bd/UKBAP-BiodiversityActionPlan-1994.pdf. Accessed 02 February 2021.

Hooke, R. L., Duque, M. J. F., & Pedraza, J. D. (2012). Land transformation by humans: A review. *GSA Today, 22*, 4–10. https://www.geosociety.org/gsatoday/archive/22/12/pdf/i1052-5173-22-12-4.pdf. Accessed 02 February 2021.

Ihrig, J., Weinbach, G. C., & Wolla, S. A. (2021). Teaching the linkage between banks and the fed: R.I.P. Money Multiplier. Econ Primer. Federal Reserve Bank of St. Louis. https://research.stlouisfed.org/publications/page1-econ/2021/09/17/teaching-the-linkage-between-banks-and-the-fed-r-i-p-money-multiplier. Accessed 02 January 2022.

IPBES. (2019). Summary for policymakers of the global assessment report on biodiversity and ecosystem services of the Intergovernmental Science-Policy Platform on Biodiversity and Ecosystem Services. https://ipbes.net/global-assessment Accessed 02 February 2021.

Jensen, M. C., & Meckling, W. H. (1976). Theory of the firm: Managerial behaviour, agency costs, and ownership structure. *Journal of Financial Economics*, 305–360. https://doi.org/10.1016/0304-405X(76)90026-X. Accessed 02 February 2021.

Krausmann, F., et al. (2017). Global socioeconomic material stocks rise 23-fold over the 20th century and require half of annual resource use. *Proceedings of the National Academy of Sciences of the United States of America, 114*(8), 1880–1885. https://doi.org/10.1073/pnas.1613773114

Lewis, J. S. (2011). "Demandite" and resources in space. John Lewis. http://www.johnslewis.com/2011/01/demandite-and-resources-in-space.html. Accessed 02 March 2022.

Lerner, S. (2015). The Teflon Toxin. The Intercept. https://theintercept.com/series/the-teflon-toxin/. Accessed 02 July 2021.

Lewis, J. S. (1996). *Mining the sky: Untold riches from the asteroids, comets, and planets*. Addison Wesley Publishing Company.

Maxwell, S. L., Fuller, R. A., Brooks, T. M., & Watson, J. E. (2016). Biodiversity: The ravages of guns, nets and bulldozers. *Nature News, 536*(7615), 143–145. https://doi.org/10.1038/536143a

McLeay, M., Radia, A., & Thomas, R. (2014). Money creation in the modern economy. Bank of England. Quarterly Bulletin. https://www.bankofengland.co.uk/-/media/boe/files/quarterly-bulletin/2014/money-creation-in-the-modern-economy. Accessed 06 June 2020.

Mittermeier, R. A., Myers, N., Thomsen, J. B., da Fonseca, G. A., & Olivieri, S. (1998). Biodiversity hotspots and major tropical wilderness areas: Approaches to setting conservation priorities. *Conservation Biology, 12*, 516–520.

Muralikrishna, I. V., & Manickam, V. (2017). Chapter fourteen—Air pollution control technologies, In I. V. Muralikrishna, & V. Valli Manickam. Environmental management (pp. 337–397). Butterworth-Heinemann.

Murphy, P. (2019). German class-action suit seeks full refund for 383,000 VW dieselgate car owners. S&P Global. https://www.spglobal.com/marketintelligence/en/news-insights/latest-news-headlines/49201920. Accessed 12 January 2021.

Myers, N. (1989). Threatened biotas: "hotspots" in tropical forests. *Environmentalist, 8*, 1–20. https://link.springer.com/article/10.1007/BF02240252. Accessed 02 February 2021.

NASA. (2019). July 20, 1969: One giant leap for mankind. NASA. https://www.nasa.gov/mission_pages/apollo/apollo11.html. Accessed 02 February 2021.

Newmark, W. D., Jenkins, C. N., Pimm, S. L., McNeally, P. B., & Halley, J. M. (2017). Targeted habitat restoration can reduce extinction rates in fragmented forests. *PNAS, 114*(36), 9635–9640. https://doi.org/10.1073/pnas.1705834114

Nguyen, T. L.T., Laratte, B., Guillaume, B., & Hua, A. (2016). Quantifying environmental externalities with a view to internalizing them in the price of products, using different monetization models. *Resources, Conservation and Recycling, 109*, 13–23. https://doi.org/10.1016/j.resconrec.2016.01.018. Accessed 02 February 2021.

OECD. (2021). Effective carbon rates 2021: Pricing carbon emissions through taxes and emissions trading. OECD Publishing. https://doi.org/10.1787/0e8e24f5-en. Accessed 12 February 2022.

OECD. (2015). Frascati manual 2015: Guidelines for collecting and reporting data on research and experimental development, the measurement of scientific, technological and innovation activities. OECD Publishing. https://doi.org/10.1787/9789264239012-en. Accessed on 02 February 2022.

OECD. (2018). Effective carbon rates 2018: Pricing carbon emissions through taxes and emissions trading. OECD Publishing.

Papazian, A. V. (2011). A product that can save a system: Public capitalisation notes. In Collected seminar papers (2011–2012) (Vol. 1). Chair for Ethics and Financial Norms. Sorbonne University. https://papers.ssrn.com/sol3/papers.cfm?abstract_id=2388043. Accessed 02 February 2022.

Papazian, A. V. (2013a). Space exploration and money mechanics: An evolutionary challenge. International Space Development Hub. https://papers.ssrn.com/sol3/papers.cfm?abstract_id=2388010 Accessed 06 June 2021.

Papazian, A. V. (2013b). Our financial imagination and the Cosmos. Cambridge Judge Business School. University of Cambridge. https://www.jbs.cam.ac.uk/insight/2013b/our-financial-imagination-and-the-cosmos/. Accessed 12 December 2021.

Papazian, A. V. (2017). The space value of money. Review of financial markets (Vol. 12, pp. 11–13). Chartered Institute for Securities and Investment. Available at: http://www.cisi.org/bookmark/web9/common/library/files/sironline/RFMJan17.pdf. Accessed 16 December 2020.

Papazian, A. V. (2021). Sustainable finance and the space value of money. Cambridge Judge Business School. University of Cambridge. https://www.jbs.cam.ac.uk/insight/2021/sustainable-finance-and-the-space-value-of-money/. Accessed 02 February 2022.

Parker, D. M. (1995). Habitat creation—A critical guide. English Nature. English Nature Science 21. http://publications.naturalengland.org.uk/publication/2294780 Accessed 02 February 2021.

Pearce, D., & Moran, D. (1994). The economic value of biodiversity. International Union for the Conservation of Nature—The World Conservation Union. Taylor and Francis.

Scott, R. T. (2018). *The universe as it really is: Earth, space*. Columbia University Press.

SETAC. (1993). Guidelines for life-cycle assessment: A "Code of Practice". Society of Environmental Toxicology and Chemistry. https://cdn.ymaws.com/www.setac.org/resource/resmgr/books/lca_archive/guidelines_for_life_cycle.pdf, Accessed 02 February 2021.

SETAC. (1991). A technical framework for life cycle assessment. Society of Environmental Toxicology and Chemistry. https://cdn.ymaws.com/www.setac.org/resource/resmgr/books/lca_archive/technical_framework.pdf. Accessed 02 February 2021.

Shepherd, P., Gillespie, B. S., & Harley, D. (1999). Preparation and presentation of habitat replacement costs estimates. English Nature. 345. http://publications.naturalengland.org.uk/publication/63034. Accessed 02 February 2021.

Shift. (2021). Leadership & governance indicators of a rights respecting culture. Shift. https://shiftproject.org/resource/lg-indicators/about-lgis/. Accessed 02 February 2022.

Shleifer, A., & Vishny, R. (1989). Managerial entrenchment: The case of firm-specific assets. *Journal of Financial Economics, 25*, 123–139.

Shleifer, A., & Vishny, R. (2012). A survey of corporate governance. *The Journal of Finance, 52*(2), 737–783.

Sisk, T. (2020). A lasting legacy: DuPont, C8 contamination and the community of Parkersburg left to grapple with the consequences. Environmental Health News. https://www.ehn.org/dupont-c8-parkersburg-2644262065.html. Accessed 02 January 2021.

Smolin, L. (2006). *The trouble with physics*. Penguin Books.

Tittensor, D. P., et al. (2014). A mid-term analysis of progress toward international biodiversity targets. *Science, 346*(6206), 241–244. https://doi.org/10.1126/science.1257484

Trucost,. (2013). *Natural capital at risk: The top 100 externalities of business*. Trucost.

UNDESA. (2021). Sustainable development goals. United Nations Department of Economic and Social Affairs. https://sdgs.un.org/goals/goal16. Accessed 12 January 2022.

US Senate. (2015). U.S. Commercial Space Launch Competitiveness Act. Public Law No: 114–90 (11/25/2015). https://www.congress.gov/bill/114th-congress/house-bill/2262. Accessed 02 December 2021.

Vallero, D. A. (2019). Chapter 8—Air pollution biogeochemistry. In: Daniel A. Vallero. *Air pollution calculations* (pp. 175–206). Elsevier. https://doi.org/10.1016/B978-0-12-814934-8.00008-9. Accessed 02 February 2021.

Van der Hoven, I. (1975). Effects of time and height on behavior of emissions. *Environmental Health Perspectives, 10*, 207–210. https://doi.org/10.1289/ehp.7510207. Accessed 02 February 2021.

Venter, O., et al. (2016). Sixteen years of change in the global terrestrial human footprint and implications for biodiversity conservation. *Nature Communications, 7*(1), 1–11. https://doi.org/10.1038/ncomms12558

Vermeulen, S., & Koziell, I. (2002). Integrating global and local values A review of biodiversity assessment. IIED. International Institute for Environment and Development. https://pubs.iied.org/sites/default/files/pdfs/migrate/9100IIED.pdf. Accessed 02 February 2021.

WEF. (2021). Global risks report 2021. https://www.weforum.org/reports/the-global-risks-report-2021.

White, B. T., Viana, L. R., Campbell, G., Elverum, C., & Bennun, L. A. (2021). Using technology to improve the management of development impacts on biodiversity. *Business Strategy and the Environment, 30*, 3502–3516. https://doi.org/10.1002/bse.2816

Withers, W. J. C. (2017). *Zero degrees: Geographies of the prime meridian.* Harvard University Press.

World Bank. (2020). The human capital index: 2020 update. The World Bank. https://openknowledge.worldbank.org/handle/10986/34432. Accessed 02 February 2021.

World Economic Forum and PwC. (2020). Nature risk rising. Available from: http://www3.weforum.org/docs/WEF_New_Nature_Economy_Report_2020.pdf.

WRAP. (2018). Gate fees 2017/18 final report: Comparing the costs of alternative waste treatment options. WRAP. http://www.wrap.org.uk/sites/files/wrap/WRAP%20Gate%20Fees%202018_exec+extended%20summary%20report_FINAL.pdf. Accessed 26 December 2018.

WWF. (2018b). Living planet report—2018b: Aiming higher. In M. Grooten, & R. E. A. Almond (Eds.), WWF. https://www.worldwildlife.org/pages/living-planet-report-2018b.

WWF. (2020). Global futures: Modelling the global economic impacts of environmental change to support policy-making. Technical Report. Available at: https://www.wwf.org.uk/globalfutures.

6

Integrating Impact into Value

C'est une question de discipline, me disait plus tard le petit prince. Quand on a terminé sa toilette du matin, il faut faire soigneusement la toilette de la planète.

Antoine de Saint-Exupéry, Le Petit Prince, 1943.[1]

How wonderful it is that nobody need wait a single moment before starting to improve the world.

Anne Frank, The Diary of a Young Girl, 1947

Having established the principle and the necessity to integrate the space impact of cash flows and assets into the equations that value those cash flows and assets, in the previous chapter we described the steps, elements, and metrics we can use to quantify the space impact of investments. This chapter brings the previous discussions together and introduces a method and logic through which we can integrate space impact into our value and return equations and apply the space value of money principle. Indeed, integrating the space impact of investments into our models is critical, if our financial value models are going to change in line with our new sustainability targets and priorities, and if financial education and theory are to reflect a new value paradigm that is in sync with our Net Zero target.

[1] "It is a question of discipline," the little prince told me later on. "When you have finished washing and dressing each morning, you must tend your planet".

A. V. Papazian, *The Space Value of Money*,
https://doi.org/10.1057/978-1-137-59489-1_6

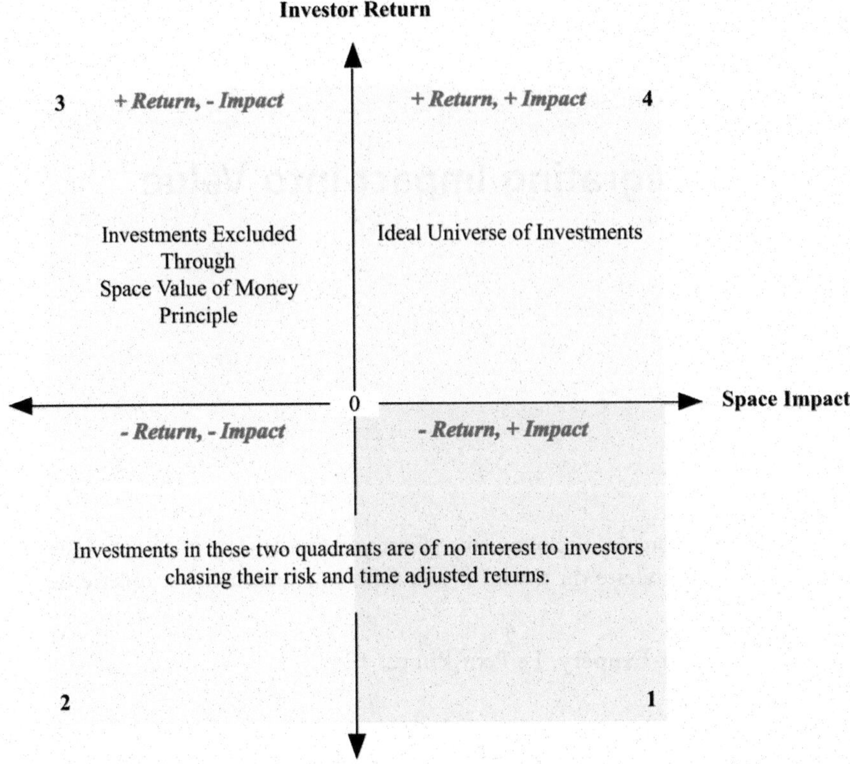

Fig. 6.1 The TRIM and the space value of money (*Source* Author)

The pollution averse planet and the aspirational human society do not impose a loss on the risk-averse investor, they simply require that the risk-averse investor selects those investments that have a positive impact on space. Thus, from the universe of all available and possible investments, the individual investor is being asked to select those that have, at the very least, no negative impact on space.

This is the transition challenge identified in chapter four. The Transition Return Impact Map Fig. 6.1 allows us to visualise the challenge and identify the impact of the space value of money principle and the associated metrics. We must prevent negative impact investments whatever their return and the space value of money provides a framework to do so.

The key change or transformation introduced through the space value of money principle is the rejection of negative impacts and the optimisation of positive impacts. The principle requires us to optimise positive space impacts as part of the value framework and establishes the bottom threshold of investment acceptability through the identification, quantification, and integration of space impacts.

Given that the space value of money principle requires that a dollar ($1) invested in space has at the very least a dollar's ($1) worth of positive impact on space, what follows is an entirely new selection mechanism.

6.1 Impact and Discounted Cash Flows

As discussed in chapter two, the main issue we have in our existing core finance models, like the NPV model and all discounted cash flow models, is that we 'treat' the actual investment with a negative sign to denote an outflow for the investor without digging into the space impact of the investment. The impact of the initial investment (II) is not included in the equation that defines the value of the future expected cash flows and is thus not part of the accept/reject decision logic of the model. Indeed, we saw how our core finance models are built in risktime, and do not account for space or the space impact of investments.

In other words, core finance equations could easily approve an investment that has a negative space impact, if the risk-adjusted return is positive for the investor. Moreover, given the omission of space and space impact, discounted cash flow models do not differentiate between an investment that will achieve an evolutionary breakthrough, like building a permanent human station on the moon, and a cement factory; or between an investment that will create new breakthrough technologies and an investment that will create none.

As discussed, and explored in detail in previous chapters, the impact of an investment becomes evident through the deployment of the initial investment. The space impact of an investment is multifaceted and multi-layered, and thus, it must be mapped before it can be quantified. Indeed, the previous chapter introduced the many aspects of space impact, and we have seen that there are three broad space impact aspects to consider: planetary, human, and economic. Each of these has its own elements and have been explored in detail in chapter five. In the following sections, I discuss an approach through which quantified impacts can be integrated into decision making and our value equations.

6.1.1 A Simplified Example

Let's assume, for example, that we are faced with a well-managed and transparent investment ($g = 1$) that has only one type of space impact, CO_2, on only one layer of space. Given that the investment is causing the pollution, this implies a negative impact. This means that the Net Space Value of the investment is given by its pollution impact.

The general equation for Pollution Impact (PI) is given as:

$$PI_{T,S,P} = \textit{Sum of All Individual Pollutant Impacts Across Years and Space Layers}$$

$$PI_{T,S,P} = \sum_{t=1}^{T} \sum_{s=1}^{S} \sum_{p=1}^{P} Q_{pst} \times C_{pst} \tag{6.1}$$

$T = \textit{Total Number of Years in the Investment}$

$P = \textit{Number/Types of Pollutants Involved}$

$S = \textit{All Space Layers Involved}$

$Q_{pst} = \textit{Quantity of Pollutant } (p) \textit{ in time } (t) \textit{ in space layer } (s)$

$C_{pst} = \textit{Cost of Cleanup for Pollutant } (p) \textit{ in time } (t) \textit{ in space layer } (s)$

Given that there is only one type of space impact, on one space layer, and only one type of pollutant, the aggregate space impact of the investment is equal to its pollution impact in terms of CO_2, and is given by the below equation:

$$NSV_{T,S} = PI_{T,S_1,CO_2} = \sum_{t=1}^{T} Q_t \times C_t \tag{6.2}$$

In this equation, Q_t is the quantity of the pollutant CO_2 across the time periods multiplied by the cost of clean-up of the pollutant C_t also across the time periods. Let's assume that the total amount of CO_2 across the years is 1000 tonnes of CO_2, spread equally across the years, 250 tonnes over 4 years.

The cost of clean-up of CO_2 emissions (CO_2 capture) is estimated at different levels by different experts and entities, and it varies between \$100 and \$1000 per tonne (Swain, 2021). Hypothetically, for the purpose of this argument, assuming that the cost of CO_2 capture is somewhere in the middle,

Table 6.1 Impacts and intensities, simplified example (*Source* Author)

	Planetary Impact	Economic Impact	Human Impact
Equation	$PI_t + BI_t$	$NAI_t + NMI_t$	$HCI_t + RDI_t$
NSV$_t$	$-125,000 + 0$	$0 + 0$	$0 + 0$
Impact intensity of CFs	$\frac{PI_t+BI_t}{Expected\ Cash\ Flow_t}$	$\frac{NAI_t+NMI_t}{Expected\ Cash\ Flow_t}$	$\frac{HCI_t+RDI_t}{Expected\ Cash\ Flow_t}$
Impact Intensity of Expected CFs	$\frac{-125,000}{100,000} = -1.25$	$\frac{0}{100,000}$	$\frac{0}{100,000}$
Sign	–	0	0

let's say it is $500 per tonne. Given that the quantity of emissions in our example is 1000 tonnes over the lifetime of the investment, this investment is actually causing an additional $500,000 of costs, on top of the initial investment, to secure the expected future cash flows.

Assuming a simple numerical example, where the investment or project involves an initial investment of $200,000 and expected yearly cash flows of $100,000 (T = 4), with a 10% discount rate, the NPV would be equal to $116,986.54, a positive value that implies the project can potentially be selected. However, this $200,000 investment is also causing an additional $500,000 negative space impact, which, based on the principle of space value of money, is not an acceptable scenario. If this cost were to be accounted for, the investment would have a negative NPV of $-383,013.46. Table 6.1 describes this simple numerical example and the elements of impact and intensities.

Thus, if the emission costs of the project were to be included in the calculation as an ecological investment necessary to generate those future expected cash flows, this would have a direct impact on the Net Present Value and would lead to a rejection of the investment. For this investment to be acceptable, it must either achieve those future expected cash flows without those emissions, or integrate a clean-up solution that recaptures the emissions cost, whatever its Net Present Value. This is based on the space value of money principle and the bottom threshold of investible projects, a Net Space Value of zero.

6.1.2 The Impact-Adjusted Present Value of Cash Flows

The previous discussion introduced, by way of a simple example, how we could consider the integration of the space impact of cash flows into the valuation of cash flows. For a general consideration of value with impact, we must integrate the impact of cash flows in a simple and effective way into our valuation exercise. The aggregate space impact of an investment has been defined as the Net Space Value of the investment, and is defined as:

$$Net\ Space\ Value_{T,S} = g \times \left(PI_{T,S,P} + BI_{T,S,B} + HCI_{T,S} + RDI_{T,S,N} + NAI_{D,S,A} + NMI_T\right) \tag{6.3}$$

$$g = Coefficient\ of\ Good\ Governance$$
$$PI_{T,S,P} = Pollution\ and\ Waste\ Impact$$
$$BI_{T,S,B} = Biodiveristy\ Impact$$
$$HCI_{T,S} = Human\ Capital\ Impact$$
$$RDI_{T,S,N} = Research\ and\ Development\ Impact$$
$$NAI_{D,S,A} = New\ Asset\ Impact$$
$$NMI_T = New\ Money\ Impact$$
$$S = All\ Space\ Layers\ Involved$$
$$T = Total\ Number\ of\ Years\ in\ the\ Investment$$
$$Net\ Space\ Value + Initial\ Investment = Gross\ Space\ Value$$

$$NSV_{T,S} + II = GSV_{T,S}$$

Figure 6.2 describes the Net Space Value of an investment as a lump sum total at the end of the space impact timeline. As we can see, alongside the future expected cash flows and the initial investment and the discount rate, we have also calculated the space impact of the investment, $NSV_{T,S}$.

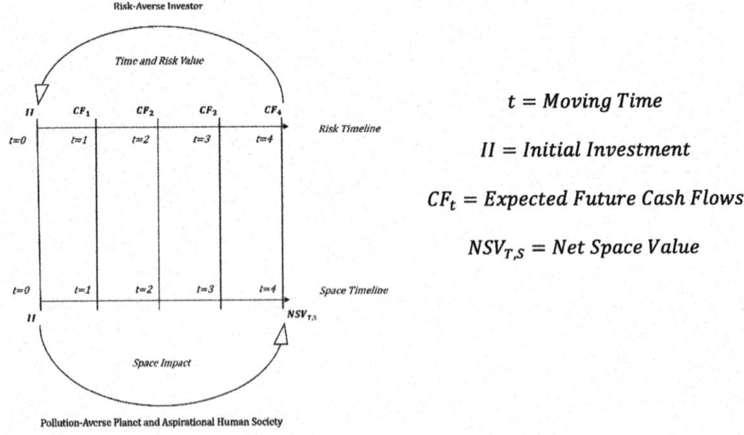

Fig. 6.2 The double timeline and net space value

In truth, the initial investment will most often be spent or used starting the first year, and thus the impact will happen as we move across the timeline. As we know from the many different space impact equations described previously, the space impact is achieved through the expenditures that constitute the investment, spread across the years as in Fig. 6.3.

This means that the space impact, $NSV_{T,S}$, can also be conceptualised as the future value of the investment and/or expenditures, as we spend the investment to earn the future expected cash flows on the risk timeline.

Given the Net Space Value of the investment, in the previous chapter we described how to derive the annualised space growth rate for the investment. We defined the space growth rate as the rate that takes us from the Initial Investment (II) in the present, to the aggregate space impact, or Net Space Value (NSV) across the space layers and T periods of the investment:

$$SPR = \sqrt[T]{\frac{NSV_{T,S}}{II}} - 1 \tag{6.5}$$

$$SPR = \textit{The Space Growth Rate per period}$$

$$II = \textit{Initial Investment}$$

$$T = \textit{Total Number of Years in the Investment}$$

$$NSV_{T,S} = \textit{Net Space Value}$$

$$\textit{Net Space Value}_{T,S} = g \times \left(PI_{T,S,P} + BI_{T,S,B} + HCI_{T,S} + RDI_{T,S,N} + NAI_{D,S,A} + NMI_T\right)$$

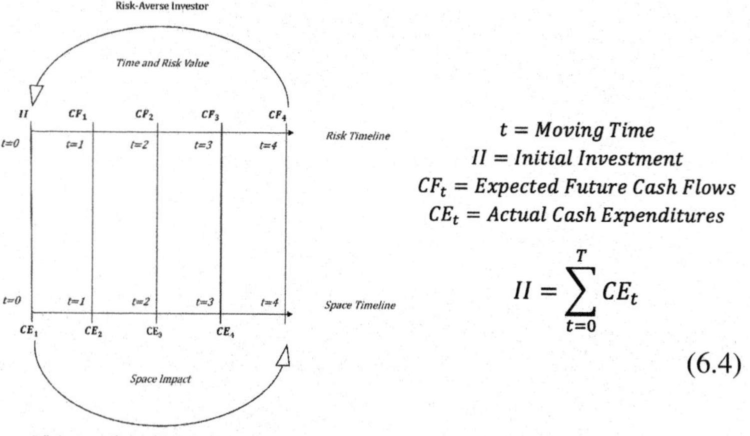

Fig. 6.3 Net space value over time

Which means that, as a summary conceptualisation, we can also write the Net Space Value as the Future Value of the investment expenditures using the space growth rate:

$$Net\ Space\ Value = NSV_{T,S}$$

$$NSV_{T,S} = \sum_{t=0}^{T} CE_t(1 + SPR)^{T-t} \tag{6.6}$$

This equation is useful to calculate the space impact of an investment using the space growth rate of the investment as a summary metric that is applied to all the expenditures involved in the investment. Moreover, this equation makes sense only for a positive space growth rate, for mathematical reasons as well as the principle of space value of money—negative space impacts must be identified and rooted out.

The concept of a *Required Space Growth Rate* is one way through which impact can be entrenched into our financial value models, such that the impact of cash flows is made integral to the equations that derive and define the value of cash flows. Requiring a 10% space growth rate on expenditure means that expenditures will have no negative planetary, human, or economic impact and will achieve a positive and growth enhancing impact on space. A 0% space growth rate implies indifference to space impact but does not involve any negative impacts.

The space value of money requires that a dollar invested in space has at the very least a dollar's worth of positive impact on space, thus, negative impacts are ruled out, and thus taken out of the investable universe. The negative impacts are identified through the quantification process of the different elements of the Net Space Value equation given below:

$$Net\ Space\ Value_{T,S} = g \times \left(PI_{T,S,P} + BI_{T,S,B} + HCI_{T,S} + RDI_{T,S,N} + NAI_{D,S,A} + NMI_T\right)$$

When an investment has negative space impacts, those negative impacts must be made to affect its value negatively in order ensure and entrench responsibility.

When an investment has negative space impacts, we adjust the value of future expected cash flows for those negative space impacts by rewriting the Net Present Value equation as follows:

$$Negative\ Impact\ Adjusted\ Net\ Present\ Value = NIA\ NPV$$

$$NIA\,NPV = -|NNSV_{T,S}| - II + \sum_{t=1}^{T} \frac{CF_t}{(1+r)^t} \tag{6.7}$$

$$NNSV_{T,S} = \textit{The Sum of Negative Impacts Across All Years and Space Layers}$$

The negative space impacts involved in the investment are summarised by the NNSV element of the equation and reflect the sum of all *negative impacts* across the years of the investment. *This is not the negative of the total NSV figure, but the sum of the negative elements within NSV*. *NNSV* identifies the negative impacts that will take to achieve those cash flows on top of the initial investment.

Potentially, given the time sensitivities of the NPV model and the risk and time conscious investor, we could also consider the *present value of the negative impacts*. We can rewrite the above by discounting the future negative impacts into the present to reveal the time value of the negative impacts using the same discount rate used for the expected cash flows:

$$-\sum_{t=1}^{T} \frac{|NNSV_t|}{(1+r)^t} = \textit{The Sum of Negative Impacts Discounted to Present} \tag{6.8}$$

Once again, the NNSV element here is the sum of the negative impacts given the time period they occur, discounted to the present at the same discount rate r. I use the absolute value and add the negative external to the total for theoretical clarity.

$$NIA\,NPV = -\sum_{t=1}^{T} \frac{|NNSV_t|}{(1+r)^t} - II + \sum_{t=1}^{T} \frac{CF_t}{(1+r)^t} \tag{6.9}$$

When investments have no negative impacts, i.e., all elements of NSV are positive, we can also consider adding the positive space impact as an optimisation target on the value of cash flows. I discuss this in the next section.

6.1.3 Impact-Adjusted Firm Value with Constant Growth

In chapter two, we discussed some of the commonly used equations used to value stocks and companies. As shown, such models or equations also rely on the discounting method. The simplest is the one that assumes a constant growth rate in the future free cash flows to the firm (FCFF) or to equity (FCFE). As introduced in chapter two, below are the constant growth equations for firm value.

$$Firm\ Value = \frac{FCFF_1}{WACC - g} \tag{6.10}$$

$$g = \sim Constant\ Growth\ Rate\ in\ FCFF$$
$$WACC = Constant\ Cost\ of\ Capital$$
$$FCFF_1 = Next\ Period\ Free\ Cash\ Flow\ to\ Firm$$

The version proposed by Koller et al., (2015, 17) is written as:

$$Firm\ Value = \frac{NOPLAT_{t=1} \times \left(1 - \frac{g}{ROIC}\right)}{WACC - g} \tag{6.11}$$

$$NOPLAT = Net\ Operating\ Profits\ Less\ Adjusted\ Taxes$$
$$g = Growth\ rate\ in\ NOPLAT$$
$$ROIC = Return\ on\ Invested\ Capital$$
$$WACC = Weighted\ Average\ Cost\ of\ Capital$$

Based on the space value of money principle, negative space impacts are not acceptable and thus must be either rectified or remedied for, and to achieve this they must affect the value of the firm. If the firm has zero space impact, a space neutral position, then the equation of firm value can remain the same. However, if the firm has negative space impacts, then its current firm value must be adjusted.

Thus, we consider the negative space impacts of the firm at t − 1 and t as our starting point.

$$NNSV_{t-1,S} = Known\ sum\ of\ negative\ impacts\ last\ year$$

$$NNSV_{t,S} = Known\ sum\ of\ negative\ impacts\ this\ year$$

Assuming that the same assumptions that apply to future expected cash flows, also apply to the space impact of the firm, i.e., they will continue to grow (higher negative value) at a constant growth rate, while remaining negative, we can quantify the space impact of the firm moving forward.

First, we define the growth rate in negative space impact as j_n.

$$j_n = \frac{|NNSV_{t,S}| - |NNSV_{t-1,S}|}{|NNSV_{t-1,S}|} \tag{6.12}$$

Treating the future negative space impacts of the firm just like the future expected free cash flows of the firm, assuming the same cost of capital, the current total value of the firm's future negative space impact can be given by the below equation:

$$\textit{Value of Negative Space Impact} = NNSV_{T,S} = -\frac{|NNSV_1|}{WACC - j_n} \tag{6.13}$$

Adjusting Eqs. (6.10) and (6.11) accordingly, we have new firm value equations that take into account the negative impacts of the future expected cash flows.

$$\textit{Neg Impact Adjusted Firm Value} = \textit{NIA Firm Value} = -\frac{|NNSV_1|}{WACC - j_n} + \frac{FCFF_1}{WACC - g} \tag{6.14}$$

$$\textit{Neg Impact Adjusted Firm Value} = -\frac{|NNSV_1|}{WACC - j_n} + \frac{NOPLAT_{t=1} \times \left(1 - \frac{g}{ROIC}\right)}{WACC - g} \tag{6.15}$$

Theoretically, if negative impacts take away value, then positive impacts should increase the value of the firm. This is very true, but *positive impacts should only be considered for firms that have no negative impacts*. This is so in order to avoid turning impact into a mathematical accounting game of arbitrage between negative and positive impacts. Given the state of the world and our ecosystem, we simply cannot afford to allow such a loophole.

In other words, if a firm has no negative impacts, then investors can take into account the positive impacts of the firm, and they can do so by adding the value of its impact (NSV) on the conventionally measured value of the firm. For firms with all positive impacts:

$$\textit{Positive Impact Adjusted Firm Value} = \textit{PIA Firm Value} = \frac{NSV_1}{WACC - J_p} + \frac{FCFF_1}{WACC - g} \tag{6.16}$$

$$Positive\ Impact\ Adjusted\ Firm\ Value = \frac{NSV_1}{WACC - J_p} + \frac{NOPLAT_{t=1} \times \left(1 - \frac{g}{ROIC}\right)}{WACC - g} \tag{6.17}$$

In Eqs. (6.16) and (6.17), NSV_1 is the total next period Net Space Value, as all elements are positive. Moreover, in the positive impact adjustment scenario, the growth rate J_p is equal to:

$$J_p = \frac{NSV_{t,S} - NSV_{t-1,S}}{NSV_{t-1,S}} \tag{6.18}$$

6.1.4 Impact-Adjusted Firm Value with Variable and Dual Period Growth

In chapter two, we also discussed firm value equations with variable growth rates, projected on a dual period basis. An initial period with variable growth rates, and another where growth is constant.

The equation that applies in this case is given as:

$$Firm\ Value = \sum_{t=1}^{T} \frac{FCFF_t}{(1 + WACC)^t} + \frac{FCFF_{T+1}}{(WACC - g).(1 + WACC)^T} \tag{6.19}$$

Given that the short-term variable window is introduced out of concern for the time sensitive investor, from a negative space impact perspective, it has no tangible difference. Thus, we can use the same expression used to adjust the constant growth model using the first period NNSV.

$$Value\ of\ Negative\ Space\ Impact = NNSV_{T,S} = -\frac{|NNSV_1|}{WACC - j_n}$$

$$Negative\ Impact\ Adjusted\ Firm\ Value = -\frac{|NNSV_1|}{WACC - j_n} + \sum_{t=1}^{T} \frac{FCFF_t}{(1 + WACC)^t} + \frac{FCFF_{T+1}}{(WACC - g).(1 + WACC)^T} \tag{6.20}$$

I use the absolute value of negative space impacts and assign the negative to the relevant sums in order to simplify and clarify the logic being applied.

When firms or investments do not comply with the space value of money principle, and have negative space impacts, in any period or in any space layer, that negative impact must directly affect their valuation, dollar for dollar, or present value of dollar impact for present value of dollar impact.

Similarly, when firms have no negative impacts, their positive impact can be taken into account by those investors who see the importance of the positive impact and the positive feedback it can have on their performance. When all impacts are positive, then the value of space impact can be added on the value of the firm for a positive impact-adjusted value.

$$\begin{aligned} \textit{Positive Impact Adjusted Firm Value} = \frac{NSV_1}{WACC - J_p} + \sum_{t=1}^{T} \frac{FCFF_t}{(1 + WACC)^t} \\ + \frac{FCFF_{T+1}}{(WACC - g).(1 + WACC)^T} \end{aligned} \tag{6.21}$$

In Eq. (6.21), NSV_1 is the first period total Net Space Value, as all elements are positive.

6.2 Impact-Adjusted CAPM, APT, and Three Factor Model

In chapter two, we explored a number of return models commonly used in finance, showing how they are abstracted from space and do not consider the impact of investments as an integral element of the return on investment. While the previous discussion looked at the NPV model and related discounted cash flow models, in this section I discuss the CAPM, APT, and Three Factor Models in the context of impact-adjusted returns.

In the CAPM model, the return on a security is conceived as a function of the return on the risk-free rate, and a risk premium for being in the security. The value principle that governs this model is risk and return, where return is defined by the level of risk, where risk is defined as past volatility vis-à-vis the market return.

$$R_i = R_f + \beta_i \times \left(R_m - R_f\right) \tag{6.22}$$

$$R_i = Return\ on\ security\ i$$
$$R_f = Risk\ Free\ Rate$$
$$\beta_i = Volatility\ measure$$
$$R_m = Return\ on\ market$$

where Beta of the security is equal to:

$$Beta_i = \beta_i = \frac{Covariance_{R_i, R_m}}{Variance_{R_m}} \tag{6.23}$$

The APT model which incorporates the CAPM proposition, and the specific multi factor model offered by Fama and French, the Three Factor Model, have similar conceptualisations with multiple factors. I use the Three Factor Model to continue the discussion.

$$E(R_i) - R_f = b_i\big(E(R_M) - R_f\big) + s_i E(SMB) + h_i E(HML) \tag{6.24}$$

where,

$$E(R_i) - R_f = Expected\ Excess\ Return\ on\ Stock\ i$$
$$E(R_i) = Expected\ Return\ on\ Stock\ i$$
$$R_f = Risk\ Free\ Rate$$
$$E(R_M) = Expected\ Return\ on\ Market$$
$$E(R_M) - R_f = Expected\ Market\ Risk\ Premium$$
$$E(SMB) = Expected\ Size\ Premium$$
$$E(HML) = Expected\ Value\ Premium$$
$$b_i,\ s_i,\ h_i = Factor\ Sensititivies\ or\ Loadings$$

Applying the space value of money principle to this model implies that the space impact of the investment must be at least neutral, as negative impact is not tolerated. In other words, the return relative to risk, size, or value can only be relevant for those investments where the space impact is neutral or positive. All those that have a negative impact should not be considered viable investments, based on the space value of money principle.

Thus, the CAPM, APT, and Three Factor Model become conditional models, where the key condition is that the Net Space Value of the investment or security is null or positive:

$$NSV_{T,S,i} \geq 0$$

$$SPR_i \geq 0$$

Given that negative space impact is not acceptable within the space value of money framework, in reality, the risk return relationship is only secondary in the investment decision tree. Thus, the assessment of risk-adjusted return makes sense for those investments where the space value condition is met.

For those investments that do have a zero or positive space impact, using the same relationship between risk and return to define the relationship between impact and return, i.e., investors can expect a higher return for additional positive impact, we need to define a benchmark return for space impact.

Let's identify *the return associated with an investment with a minimum space impact* as R_{min}. This is the return associated with an investment that has the lowest tolerated space impact within the space value of money framework. While theoretically this is zero (NSV = 0), in practice the benchmark could be higher.

$$R_{min} = R_{minimum} = R_{investment\ with\ minimum\ tolerated\ space\ impact}$$

What investment or instrument could be used to reflect a minimum space impact investment? The zero or minimum space impact rate could be the return on a government bond or a public investment. However, the space impact of public investments are rarely zero. Moreover, this is not always straightforward to calculate given the diversity of public investments and given that the proceeds of a government bond can and are usually used to fund expenditures and a diverse set of public projects, which do not have a uniform impact in terms of waste and pollution, biodiversity, asset creation, etc.

However, for conceptual simplicity, a secondary government bond transaction comes closest to a reliable return on an investment with minimum space impact. Thus, we can assume that the return associated with the minimum space impact investment is the return on the government T-bill, which is also the risk-free rate. Indeed, governments should set the lower threshold of positive space impact, becoming a benchmark of responsibility—ideally higher than zero space impact.

$$R_{min} = R_{investment\ with\ minimum\ tolerated\ space\ impact} = Risk\ Free\ Rate = R_f$$

Let's also identify the return on the investment as R_i, and let us define MSP as the Market Space Premium.

$$R_i = Return\ on\ Investment\ or\ Security\ i$$

$$MSP = Market\ Space\ Premium$$

The market space premium is the expected excess return due to a higher than zero (or benchmark) positive space impact by an investment or security. In other words, it is the reward given to those companies that have a higher positive impact than the minimum threshold required by the principle of space value of money (zero).

MSP is a function of a number of factors that are company specific and market specific. The company specific aspects are linked to the type of space impact involved. At different points in time, different types of impacts may be more popular than others, or more in demand than others. Moreover, some types of impact may be seen as a source of projected future earnings, like, for example, a carbon capture technology company may be more valuable than another highly profitable company with a high space impact in the software industry.

The systemic aspects that can influence the space premium are linked to market demographics and regulations. In a market where there are regulatory guidelines and rules in relation to space impact and responsibility, investors would naturally have to work within those guidelines. Ultimately, such premiums may vary given market awareness, commitment, and culture.

The best approximation for the market space premium will be revealed over time, and once such a transformation has been adopted and integrated of course. However, theoretically, we can expect that some form of benchmarking would play a role in determining the space premium on the minimum space impact investment rate, i.e., R_f. To integrate impact as a factor in the CAPM or APT or 3FM (Three Factor Model), we also need to identify a portfolio of high impact stocks, let's call the expected return on a high impact portfolio, measured through a high NSV and high space growth rate:

$$R_{HIP} = Expected\ Return\ on\ High\ Impact\ Portfolio$$

Before we can calculate the expected space premium, an additional sensitivity measure must be introduced. I have defined this as Theta, θ, the space impact sensitivity of returns.

$$\theta = Theta = Space\ Impact\ Sensitivity\ of\ Returns$$

$$MSP = Market\ Space\ Premium = R_{HIP} - R_{min} = R_{HIP} - R_f$$

$$MSP_i = Market\ Space\ Premium\ for\ Security\ i = \theta_i.(R_{HIP} - R_f) \quad (6.25)$$

In this equation, Theta, is the historical relationship between the investment's past Net Space Values, or space impact, and its past returns.

$$\theta_i = \frac{Covariance_{R_i,\ NSV_i}}{Variance_{NSV_i}} \quad (6.26)$$

$$NSV_i = Net\ Space\ Value\ of\ Security\ i$$

To formalise the impact relationship of the return on a specific security, we can write:

$$R_i = R_f + \theta_i(R_{HIP} - R_f) \quad (6.27)$$

If we were to introduce impact as factor, not just as a condition for universe selection, then we can write:

$$R_i = R_f + \beta_i(R_m - R_f) + \theta_i(R_{HIP} - R_f) \quad (6.28)$$

$$E(R_i) - R_f = b_1(E(R_M) - R_f) + s_i E(SMB) + h_i E(HML) + \theta_i(R_{HIP} - R_f) \quad (6.29)$$

The actual relationship between positive space impact and return is something that will be tested in the future, once we have enough historical data that reveals the difference and fluctuations in return given fluctuations in space impact. Measuring such a sensitivity today may naturally result in insignificant results. When the principle of space value of money is adopted, we may be able to fine-tune our understanding of how much more investors can require or expect to earn, for each additional dollar of positive space impact, and how that affects the relationship with risk. Only then we would be able to discover the impact sensitivity of prices and returns for individual securities.

6.3 Impact-Adjusted Returns

One of the largest segments of academic and industry literature in finance concerns the search for abnormal risk-adjusted returns. Indeed, as discussed in chapter two, the search for stock market anomalies that can deliver abnormal risk-adjusted returns has covered all and every possible aspect of stocks and companies. Academics and investors alike have been on a search for the data points and correlations that could allow investors to predict the movement in stock prices, thus helping them achieve higher returns than warranted by the risk levels involved.

What we cannot find in the finance literature is the search for impact-adjusted returns. This has been left in a blind spot for the same reasons discussed in the earlier chapters of this book. It is only recently that environmental, social, and governance factors have become relevant, and till now, the frameworks being discussed do not actually involve the measurement of impact and the integration of that impact into our valuation models.

Impact-adjusted returns can be a good indicator of market integrity. Indeed, given the currently ongoing changes in the finance industry, the new suggested framework, the new concepts and tools of realignment, impact-adjusted returns can reveal how serious and how determined market participants are, and how ready they are to penalise those companies that have a negative impact, and how willing they are to reward those with a positive impact.

Today, given the variable frameworks and past patterns, we may find that stock returns are higher for those companies with higher negative impact rather than positive impact, as those would be the ones extracting the highest value out of our ecosystem without taking responsibility for the cost of their impact.[2]

The typical and previously established relationship between privatised profits and socialised costs, which has characterised the last many decades,

---

[2] Important to note here that impact is not equivalent to ESG scores, where having a policy on X or Y could positively affect your score, where lack of media coverage about controversies earns you good points, etc. These scores and their limitations have been discussed in chapter three and given that they score perception management rather than actual impact, it is not surprising that ESG scores have been found to correlate with positive market returns.

is the result of a framework serving the risk-averse investor, without any responsibility for impact. As such, searching for impact-adjusted returns to identify their relationship with risk-adjusted returns could be a promising new research area. Similarly, revisiting the market efficiency literature in the light of impact-risk-adjusted returns could yield very interesting insights.

6.4 Conclusion

This chapter explored the space value of money concepts and metrics by integrating them into a select number of core equations or models of value and return in finance. The framework and metrics have shown that it is feasible and mathematically possible to integrate impact into our value models in finance. The key is the willingness to accept the principle of space impact responsibility and the transformation it introduces.

The space value of money principle establishes the bottom threshold of investment acceptability through the identification, quantification, and integration of space impacts. It requires us to reject negative impacts and optimise positive space impacts, both critical to achieving the transition to Net Zero by mid-century.

We are facing one of the most challenging collective action dilemmas in human history. Given the fact that we have littered every aspect of our physical context, i.e., air, oceans, land, and outer space, the transformation needed is systemic and requires a paradigm shift. For a successful transition across industries, value chains, and the world, we need a new set of financial value principles and equations. The space value of money principle and metrics provide one such framework through which we can facilitate this change.

References

Brealey, A. R., Myers, C. S., & Allen, F. (2020). *Principles of corporate finance* (13th ed.). New York: McGraw Hill.

De Saint-Exupéry, A. (1943). Le Petit Prince. Reynal & Hitchcock.

Fama, E. F. (1970). Efficient capital markets: A review of theory and empirical work. *The Journal of Finance*, *25*, 383–417. https://doi.org/10.2307/2325486. Accessed 02 February 2021.

Fama, E. F. (2004). The capital asset pricing model: Theory and evidence. *Journal of Economic Perspectives*, *18*, 25–46. https://www.aeaweb.org/articles?id=10.1257/089533004216243O. Accessed 02 February 2021.

Fama, E. F., & French, K. R. (1992). The cross-section of expected stock returns. *The Journal of Finance, 47*, 427–465. https://doi.org/10.2307/2329112. Accessed 02 February 2021.

Fama, E. F., & French, K. R. (1993). Common risk factors in the returns on stocks and bonds. *The Journal of Financial Economics, 33*, 3–56. https://doi.org/10.1016/0304-405X(93)90023-5. Accessed 02 February 2021.

Fama, E. F., & French, K. R. (2015). A five-factor asset pricing model. *Journal of Financial Economics, 116*, 1–22. https://doi.org/10.1016/j.jfineco.2014.10.010. Accessed 02 February 2021.

Fibonacci, Leonardo of Pisa (1202) Liber Abaci. Translated by Sigler, LE 2002, *Fibonacci's Liber Abaci*, Springer.

Frank, A. (1952). Anne Frank: The diary of a young girl. Doubleday & Company, Inc.

Gordon, J. R., & Gordon, M. J. (1997). The finite horizon expected return model. *Financial Analysts Journal, 53*, 52–61. https://doi.org/10.2469/faj.v53.n3.2084. Accessed 02 February 2021.

Gordon, M. J., & Shapiro, E. (1956). Capital equipment analysis: The required rate of profit. *Management Science, 3*, 102–110. https://www.jstor.org/stable/2627177. Accessed 02 February 2021.

Graham, J., & Harvey, C. (2002). How CFOs make capital budgeting and capital structure decisions. *Journal of Applied Corporate Finance, 15*, 8–23. https://doi.org/10.1111/j.1745-6622.2002.tb00337.x. Accessed 02 February 2021.

Koller, T., Dobbs, R., Huyett, B., & McKinsey and Company. (2011). Value: The four cornerstones of corporate finance (6th ed.). Wiley.

Koller, T., Goedhart, M., Wessels, D., & McKinsey and Company. (2015). Valuation: Measuring and managing the value of companies (6th ed.). Wiley.

Lintner, J. (1965). The valuation of risk assets and the selection of risky investments in stock portfolios and capital budgets. *The Review of Economics and Statistics, 47*, 13–37. https://doi.org/10.2307/1924119. Accessed 02 February 2021.

Merton, R. (1973). An intertemporal capital asset pricing model. *Econometrica, 41*, 867–887. https://doi.org/10.2307/1913811. Accessed 02 February 2021.

Modigliani, F., & Miller, M. H. (1958a). Corporate income taxes and the cost of capital: A correction. *The American Economic Review, 53*, 433–443. https://www.jstor.org/stable/1809167. Accessed 02 February 2021.

Modigliani, F., & Miller, M. H. (1958b). The cost of capital, corporation finance and the theory of investment. *The American Economic Review, 48*, 261–297. https://www.jstor.org/stable/1809766. Accessed 02 February 2021.

Papazian, A. V. (2017). The space value of money. Review of financial markets (Vol. 12, pp. 11–13). Chartered Institute for Securities and Investment. Available at: http://www.cisi.org/bookmark/web9/common/library/files/sironline/RFMJan17.pdf. Accessed 16 December 2020.

Papazian, A. V. (2021). Sustainable finance and the space value of money. Cambridge Judge Business School. University of Cambridge. https://www.jbs.cam.ac.uk/insight/2021/sustainable-finance-and-the-space-value-of-money/. Accessed 02 February 2022.

Ross, S. A. (1976). The arbitrage theory of capital asset pricing. *Journal of Economic Theory, 13*, 341–360. https://doi.org/10.1016/0022-0531(76)90046-6. Accessed 02 February 2021.

Ross, S. A. (1978). The current status of the Capital Asset Pricing Model (CAPM). *Journal of Finance, 33*, 885–890. https://doi.org/10.2307/2326486. Accessed 02 February 2021.

Sharpe, W. F. (1964). Capital asset prices: A theory of market equilibrium under conditions of risk. *Journal of Finance, 19*, 425–442. https://doi.org/10.1111/j.1540-6261.1964.tb02865.x. Accessed 02 February 2021.

Sharpe, W. F. (1994). The Sharpe ratio. *The Journal of Portfolio Management, 21*, 49–58. https://doi.org/10.3905/jpm.1994.409501. Accessed 02 February 2021.

Swain, F. (2021). The device that reverses CO_2 emissions. BBC Future Planet. https://www.bbc.com/future/article/20210310-the-trillion-dollar-plan-to-capture-co2. Accessed 02 February 2022.

TCFD. (2017). Final report: Recommendations of the task force on climate-related financial disclosures. https://assets.bbhub.io/company/sites/60/2020/10/FINAL-2017-TCFD-Report-11052018.pdf. Accessed 02 February 2021.

TCFD-PAT. (2020). Measuring portfolio alignment. https://www.tcfdhub.org/wp-content/uploads/2020/10/PAT-Report-20201109-Final.pdf. Accessed 02 February 2022.

TCFD-PAT. (2021). Measuring portfolio alignment: Technical considerations. https://www.tcfdhub.org/wp-content/uploads/2021/10/PAT_Measuring_Portfolio_Alignment_Technical_Considerations.pdf. Accessed 12 December 2021.

7

The Algorithms of Sustainable Finance

So this is an algorithm. The modern meaning for algorithm is quite similar to that of recipe, process, method, technique, procedure, routine, rigmarole, except that the word 'algorithm' connotes something just a little different. Besides merely being a finite set of rules that gives a sequence of operations for solving a specific type of problem, an algorithm has five important features…: 1) Finiteness, 2) Definiteness, 3) Input, 4) Output, 5) Effectiveness.

Donald Ervin Knuth, The Art of Computer Programming, 1968

A man provided with paper, pencil, and rubber, and subject to strict discipline, is in effect a universal machine.

Alan Turing, Intelligent Machinery: A report, 1948

Digitisation has been gaining momentum over the last many decades. This ongoing transformation has triggered the architectural and technological revamp of entire value chains across industries and markets. The digital world has become a unique and seemingly irreplaceable platform through which we create, communicate, exchange, and deliver value across all economic activities, in both private and public spheres. While this software and computer driven world may be relatively new, technology has been transforming finance and industry for some time. John Maynard Keynes, writing in 1920, describes an image of global trade and finance that is not far away from our own, except for the type of technological medium used.

The inhabitant of London could order by telephone, sipping his morning tea in bed, the various products of the whole earth, in such quantity as he

A. V. Papazian, *The Space Value of Money*,
https://doi.org/10.1057/978-1-137-59489-1_7

> might see fit, and reasonably expect their early delivery upon his doorstep; he could at the same moment and by the same means adventure his wealth in the natural resources and new enterprises of any quarter of the world, and share, without exertion or even trouble, in their prospective fruits and advantages; or he could decide to couple the security of his fortunes with the good faith of the townspeople of any substantial municipality in any continent that fancy or information might recommend. (Keynes, 1920, 11)

Although fintech (financial technology) has gained special attention recently, technology has been continuously intertwined with the growth and development of banks, money, and financial markets over centuries (Arner et al., 2015). More recently, since the 1971 creation of Nasdaq in the US and the 1986 Big Bang that transformed banking and markets in the UK, technology has been integral to financial development and deepening (Clemons & Weber, 1990).

What characterises the new digital age in finance, more than any other of its many features, is the rise of Algorithmic Trading (AT) and High Frequency Trading (HFT). On different levels of complexity, they refer to electronic trading and investment strategies devised by humans and executed by machines, and they have been gaining market share since the early 2000s. In some markets, their share is around 70% of total orders (SEC, 2020; ECB, 2021; SIFMA, 2019; Gomber et al., 2011).

This growth in algorithmic trading has been fuelled by technological developments that have enhanced infrastructures, enabling fast digital communications between exchanges, market makers, brokers, and traders. This has been further enhanced through the rise of artificial intelligence and machine learning, allowing the growth of both data creation and analysis, and far more complex algorithms.

Since the Paris Agreement (UNFCCC, 2015) and the rise to prominence of sustainability and sustainable finance, or ESG integration, the finance industry has had to deal with an entirely new and very different kind of transformation. In a matter of few years, the 'rewiring' of finance is on the agenda (Herman, 2021) and the industry is faced with the necessity to make sense of the digital and sustainability transformations simultaneously. This has revealed a gap between the existing technological toolkit of the industry and its new sustainability frameworks and targets—a market opportunity that many are keen to develop.

Meanwhile, given the digital context, digitisation has become a key requirement to sustain and maintain the growth of sustainable finance. This has led to an exponential growth in sustainable tech solutions in finance. Tools that help companies measure and report their emissions, ESG performance, and compliance, applications that help asset managers screen public equities, platforms that provide impact measurement and reporting ecosystems are some examples.

The necessity of digitisation in sustainable finance is existential. In the age of algorithmic trading, where global capital markets are driven by technology, and digital tools have automated strategies and trading for decades already, expanding market share and reeling in a greater pool of capital will require equally efficient and fast technological solutions.

7.1 Digitising Confusion

The existential necessity to digitise sustainable finance is primarily challenged by the lack of theoretical clarity and the variability of the frameworks and tools currently used within the field. We have seen in chapter three that the existing temperature scores are indicators of misalignment rather than tools of transformation, and that the current TCFD framework is a variable and voluntary framework with many moving parts left to the discretion of those applying the framework. We also discussed the many challenges of ESG ratings, which are infrequent and incomparable opinion points that fall short of providing market participants with an analytical framework that can be used to interpret and act upon new information that happens to reach the market in between rating updates. Indeed, chapter three demonstrated that the standards, frameworks, and alignment tools in the sustainable finance field do not go far enough to transform our principles and equations of value in core finance theory and practice.

When the equations of value are yet to be defined, and we are still faced with infrequent and incomparable opinion points or general misalignment indicators like temperature scores, building decision-making tools that can integrate responsibility into finance and entrench sustainability into our capital markets is a challenging process.

In the preface of this book, I started the discussion by pointing out that digitisation does not automatically imply improvement, that an unfair and unequal process can remain so after digitisation. Indeed, the architecture of our markets and the principles and equations of a valuation model change not when they are digitised, but when they are reinterpreted in their fundamental assumptions, and are rebuilt upon an entirely new value framework.

In the age of digital transformation, we are more than ever exposed to the risk of digitising confusion, and reinforcing suboptimal frameworks, structures, and models. Today, this risk is highest in the field of sustainable finance.

Moreover, digitisation does not automatically imply effectiveness. In the case of sustainable finance, given what is at stake, digitising incomplete and ineffective frameworks can delay the transition and have serious implications for everyone involved.

As Knuth (1968, 5) describes, the first two main features of an algorithm are (1) finiteness, and (2) definiteness. Using Knuth's own words, finiteness refers to the fact that '[a]n algorithm must always terminate after a finite number of steps'. In other words, an algorithm is not an open-ended process. While some can add or create new decision loops within themselves, even those are finite as they are predefined. According to Knuth, definiteness refers to the fact that '[e]ach step of an algorithm must be precisely defined; the actions to be carried out must be rigorously and unambiguously specified for each case'.

Taking these key features into account, it is safe to say that it is only after investors are provided with a clear and transformed value framework that effective sustainable finance algorithms will become possible. Given the already advanced digital landscape of capital markets, those algorithms will be key to maintain, preserve, and expand the market reach of sustainable finance.

Investors need a reliable and replicable framework that can help them interpret new information relevant to the sustainability of their investments. By providing such a framework, we help the development of new equations of value with new data points, which will in turn create the opportunity to build robust decision-making algorithms that can entrench sustainability into an already digitised market.

Going back to Knuth's work (1968, 6), he describes Inputs and Outputs as the third and fourth key features of an algorithm. Quoting Knuth directly:

> (3) *Input*. An algorithm has zero or more *inputs:* quantities that are given to it initially before the algorithm begins, or dynamically as the algorithm runs. These inputs are taken from specified sets of objects.... 4) *Output*. An algorithm has one or more *outputs:* quantities that have a specified relation to the inputs (Knuth, 1968, 6).

While algorithms have grown in complexity, their basic building blocks remain the same. The inputs are the data points or objects that are fed into the algorithm, and they are central for any output to be produced based on the steps and specific logic of the algorithm. In sustainable finance trading algorithms, outputs are often information based buy and sell or hold decisions. Which brings us back to the principles, equations, and data points we must clarify for effective sustainable finance algorithms. Indeed, effectiveness is the fifth key feature described by Knuth (1968, 6).

To clarify, the argument here refers to using algorithms in sustainable finance and trading. I am not discussing the role of Artificial Intelligence (AI) and other machine learning tools in making our value chains more sustainable, in business or industry. Indeed, the digital transformation and the sustainability transformation that we are going through are relevant to each other and to our daily lives. Achieving our sustainability targets and objectives in the digital age will very much depend on effective digital and AI tools.

The growth in spatial finance discussed in chapter two is one example. Using AI, machine learning and geospatial data to analyse investments and their operations from a sustainability perspective will become ever more central to the transition and to sustainable finance. Indeed, the use and importance of geospatial data and natural language processing (NLP) is already growing. Many companies use them for land management purposes, and ESG ratings providers use them to flag risks in texts or data. However, although these tools support data creation and observation, and act as research assistants to analysts, they do not and cannot provide the market with new decision-making frameworks, and new value equations that integrate impact into our models, facilitating sustainable investment and trading algorithms. This has to be achieved in finance theory and practice first, through a transformed mathematics of value and return.

As we crystallise the principles, models, equations, and data points through which we will entrench sustainability in the theory and practice of finance, we will undoubtedly move from ESG ratings and Temperature scores to Sustainable Finance Algorithms—involving more than engagement, built on a more robust decision tree, and covering our multifaceted impact on space and its many layers. In the next section, I explore some of the key questions and challenges that we must address in order to create and deploy the sustainable finance algorithms of the future.

7.2 Digitising Sustainability

7.2.1 Data

We have all read or heard the expression 'the age of Big Data is here', and we have all been exposed to the changes this new age has triggered in all aspects of our lives, from business to finance and industry (Laney, 2001; Carter, 2011;

Boyd & Crawford, 2012; Lohr, 2012; Hartford, 2014; McKinsey, 2011, 2016).

The growth of data is the first and most immediate effect of digitisation. David Spiegelhalter describes the challenges with big data eloquently, he says: '[t]here are a lot of small data problems that occur in big data, [t]hey don't disappear because you've got lots of the stuff. They get worse' (quoted in Hartford, 2014). Indeed, the challenges of Big Data get more daunting when one has no clear framework of analysis (Corea, 2019; Doornik & Hendry, 2015).

It is commonly discussed and pointed out that one of the main challenges facing sustainable finance, and specifically the digitisation of sustainable finance, is the lack of data. Given the growing demand for such information, and the sustainable finance transformation, this is rightfully considered a huge market opportunity. As Blanque (2021) says, 'the battle for data has just begun'.

While the drive to clarify and create the data needed for the new sustainable investing landscape intensifies, many of the large public corporations listed on major exchanges and included in key indices do not yet report their emissions. Given a variable and voluntary framework and the birth of new frameworks still in development, the data challenges of sustainable finance will most likely be with us for the foreseeable future.

The first general truth about data, which must be stated for conceptual and theoretical clarity, is that data always describes the past. No matter how recent the data, a nanosecond ago or even a femtosecond (10^{-15} of a second) ago, data is the past. In truth, data is like ash; it can only exist after the wood has burnt and the femtosecond has passed. My car's GPS data can tell you where I have been, or where I am at this moment a second or so ago, but not where I may go tomorrow. Moreover, even when we find and reveal repeating patterns in past data, their repetition in the future is at best a probabilistic expectation.

The second general truth about data[1] is that its ultimate source is human imagination and action. Indeed, tomorrow early morning I may decide to visit Merlin's cave beneath Tintagel Castle, or I may not. Alternatively, I may go somewhere I have never been before, inspired by a person I have not met yet. Or, I may imagine going to Merlin's cave, but decide not to act on it. Thus, while imagination is the ultimate source of data, the creation of data occurs when imagination takes action, and thus makes observable

[1] Here, I refer to data generated through human activities. Some data, though observed and recorded by humans, are not about human activities, like the movements and orbits of celestial objects.

changes in the physical world, interacting digitally or non-digitally with the environment.

The GPS data of my trip to Cornwall will never exist if I end up changing my mind and will leave a trail of digital and non-digital data if I decide to go. If I stop for gas on the way, the CCTV camera at the station will record me at the pump, if I pay with my debit card, then my statement will reveal the transaction, etc. When we interact with the world digitally, or through the use of products and services that have integrated digital interfaces, we automatically create digital data. Structured or unstructured, clean or unclean, we leave a digital footprint.

When we interact with the world without any digital intermediation, like walking to the beach without any digital product or service, we may create non-digital data. For example, by observing the plastic bottles on the seashore, counting them, and recording the number, date, and time, I create the data point. If, however, I do not observe the plastic bottles on the beach, the data point 'the number of plastic bottles on day x on beach y at time t' is not created. This does not mean that there were no plastic bottles on the beach, it just means the observation was not made. If I observe the plastic bottles but fail to record the observation, the non-digital data becomes possible, but must be recorded for it to exist as a data point. When we are interacting with the non-digital world, the starting point of data creation is observation followed by the recording of that observation in one shape or form.

Figure 7.1 provides a schematic representation of some of the key aspects

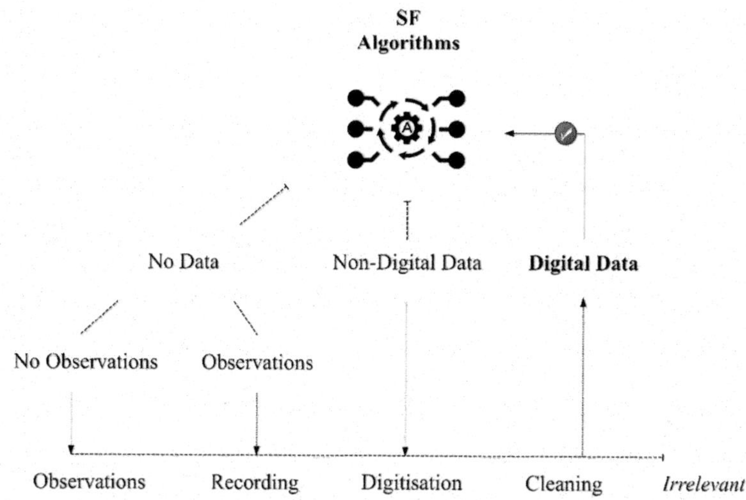

Fig. 7.1 Data requirements: sustainable finance algorithms (*Source* Author)

to take into account as we make further progress in developing the required data for sustainable finance. The first challenge is to identify whether the data exists—in digital or non-digital form. If the data does not yet exist, then we must ask ourselves the next question, the data does not exist because the observations do not exist, or because the observations exist, but have never been recorded. The fact that companies do not report their emissions does not automatically mean that they have not observed their emissions, and if they have observed them, it does not automatically mean that they have recorded the observations.

Given the past focus on risk and time, and the fact that climate and sustainability became mainstream concerns only recently, companies have never had the need or the necessity to calculate their emissions or their pollution and biodiversity impact the way they had to report their quarterly profits. Markets and investors did not ask for them. In other words, the sustainable data creation process for many entities will often begin at the most basic level of observation.

The other type of data that also presents some challenges is data that exists but in non-digital form. In other words, observations that have been recorded but not structured nor translated into a digital format or database. Although new and relatively young institutions and companies may not have any significant non-digital data, the case of governments and older firms is different. Thus, we could also be faced with the need to translate sustainability records from analogue to digital technology before such data become usable.

Last but not least, existing digital data presents its own set of challenges. First, just because the data is in digital form it does not mean that it is relevant, and second, when it is relevant, it does not mean it is structured and clean data. In other words, digital data could be available but incomplete, filled with errors, and unstructured.

These challenges must be taken into account as we define the frameworks and data points we need for an effective transition and digitisation of sustainable finance. Given the many equations and elements of impact discussed in this book, the data we need to measure the pollution, biodiversity, human capital, R and D, new asset, and new money impact of an investment may need to be required, observed, digitally recorded, structured, and reported accordingly.

7.2.2 Algorithms

The technical challenges regarding data availability and quality in sustainable finance are directly linked to the lack of a crystallised value paradigm that transforms our existing equations of value and return. The new data points we need will be ultimately defined through the new framework and equations. The purpose and main contribution of this book is to propose an approach that could fill the theoretical gap that exists between core finance theory and sustainable finance.[2]

Indeed, the principle of space value of money and the many equations of impact proposed and discussed provide a possible framework for the digitisation of sustainable finance. Once we have adopted the space value of money principle and relevant equations, the data points that we need become self-evident. Once we have the data points identified we will be able to build and apply decision tools that can take action, whether in terms of public trading or private investments—thus digitising responsibility and integrating sustainability in our markets without the intermediation of scores.

Figure 7.2 describes the algorithmic process applied to potential sustainable finance trading tools. Today, the algorithmic trading tools available in the market, besides those focused entirely on technical analysis, or purely volume and price analysis, utilise the existing value framework within finance, built on risk and return and time value of money, using different existing anomalies or trading strategies that are not linked to the multi layered space impact of stocks.

Although some asset managers and traders use ESG ratings or scores to track and compare specific stocks and portfolios, these scores do not have the frequency and transparency to become active decision-making elements in trading algorithms. ESG scores, as discussed in chapter three, are interpretations of data that the market cannot always decipher. Moreover, the frequency of release of ESG scores does not compare with the frequency of data and information in conventional capital markets.

Furthermore, even when reported and updated more often, an ESG score does not provide a logical and clear method of evaluating new information and/or news that affect the company and its ESG or sustainability performance. There is no framework that helps investors interpret marginal information in between rating updates, and in a market that acts and reacts in nanoseconds (Markoff, 2018), this is a significant obstacle for wider adoption and consideration.

The argument here is not about interpretations. Interpretations are everywhere, and even in the space value framework proposed in this book, the many different coefficients involved require investors to make their own interpretive assessment of different features of the investment. Without divergent

[2] The lack of emissions data is due to the fact that they have only recently been identified as material information, and the fact that the TCFD framework is still a voluntary one.

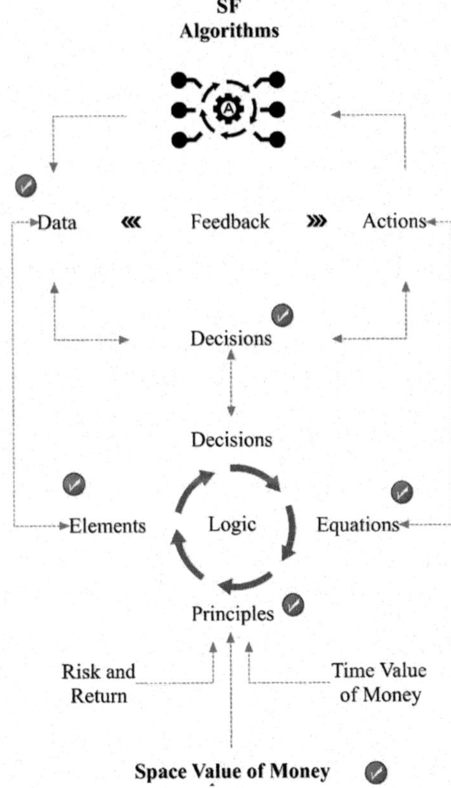

Fig. 7.2 Principles of finance, sustainability, and SF algorithms (*Source* Author)

interpretations there would be no market. The point here is that ESG scores do not provide a transparent theoretical framework through which investors can make their own event-based or news-based assessments.

Investors and traders react to news or understand how others will react to a specific action or event based on existing value relationships built within the discipline and market practices. ESG ratings and engagement focused temperature scores do not provide such an interpretive framework.

To provide algorithmic trading and human traders the ability to interpret new sustainability relevant information, there must be a set of clear principles, equations, and related data points upon which decisions can be made, and actions can be taken by the market.

7.3 Conclusion

Digitisation does not automatically imply improvement, nor does it imply effectiveness. Given the theoretical background in core finance theory (chapter two), and the current state of sustainable finance frameworks, tools, and scores (chapter three), we must recognise that the development of effective digital tools in sustainable finance can only come about after we have reinvented our value paradigm and entrenched responsibility into finance theory and practice.

Throughout the many chapters of this book, I proposed and argued for the introduction of the space value of money principle and the associated metrics into our analytical framework and equations of value. The purpose of the discussion was to entrench sustainability into the very core of the discipline and to provide a transformed financial value framework that integrates space and our responsibility of impact.

The principle, equations, and elements discussed in chapter four, five, and six provided a possible framework through which we could digitise responsibility, and transform the algorithms that make up and influence much of our capital markets today.

Digitisation without theoretical and methodological clarity can be counterproductive, delaying the changes necessary to achieve an effective transition. While the application of machine learning tools to geospatial data has proven to be a valuable resource, the digitisation of sustainable finance will require a structured transformation in our value paradigm, principles, and equations, so that we can clarify and identify the data points needed for the creation and deployment of sustainable finance algorithms. Indeed, in a market increasingly dominated by algorithmic and high frequency trading, expanding the pool of sustainable assets, or assets managed according to sustainable strategies, may very well depend on such developments.

References

Arner, D., Barbaris, J., & Buckley, R. (2015). The evolution of Fintech: A new post-crisis paradigm. University of Hong Kong Faculty of Law Research Paper No. 2015/047. https://www.researchgate.net/publication/313365410_The_Evolution_of_Fintech_A_New_Post-Crisis_Paradigm. Accessed 02 February 2021.

Blanque, P. (2021). The battle for data has just begun. ESG Investor. https://www.esginvestor.net/the-battle-for-data-has-just-begun/. Accessed 02 February 2022.

Bollier, D. (2010). The promise and peril of big data. Aspen Institute. https://www.aspeninstitute.org/wp-content/uploads/files/content/docs/pubs/The_Promise_and_Peril_of_Big_Data.pdf. Accessed 02 February 2020.

Boyd, D., & Crawford, K. (2012). Critical questions for big data. *Information Communication & Society, 15*(5), 662–679. https://doi.org/10.1080/1369118X.2012.678878

Carter, P. (2011). Big Data analytics: Future architectures, skills and roadmaps for the CIO. White Paper IDC.

Clemons, E. K., & Weber, B. W. (1990). London's Big Bang: A case study of information technology, competitive impact, and organizational change. *Journal of Management Information Systems, 6*(4), 41–60.

Corea, F. (2019). An introduction to data. Springer Nature.

Doornik, J. A., & Hendry, D. F. (2015). Statistical model selection with "Big Data". *Cogent Economics and Finance, 3*(1). https://doi.org/10.1080/23322039.2015.1045216. Accessed 02 February 2021.

ECB. (2021). Algorithmic trading: Trends and existing regulation. European Central Bank. https://www.bankingsupervision.europa.eu/press/publications/newsletter/2019/html/ssm.nl190213_5.en.html. Accessed 02 February 2022.

Gomber, P., Arndt, B., Lutat, M., & Uhle, T. (2011). High-frequency trading. Goethe Universitat Frankfurt AM Main. https://papers.ssrn.com/sol3/papers.cfm?abstract_id=1858626. Accessed 02 February 2021.

Hartford, T. (2014). Big data: Are we making a big mistake? *Financial Times*. https://www.ft.com/content/21a6e7d8-b479-11e3-a09a-00144feabdc0

Herman, Y. (2021). Britain's Sunak pledges to 'rewire' global finance system for net zero. Reuters News. https://www.reuters.com/business/cop/britains-sunak-pledges-rewire-global-finance-system-net-zero-2021-11-03/. Accessed 14 December 2021.

Keynes, J. M. (1920). *Economic consequences of the peace.* Harcourt.

Knuth, D. E. (1968). The art of computing volume 1: Fundamental algorithms. Addison-Wesley.

Laney, D. (2001). 3D data management: Controlling data volume, velocity and variety. META Group Research Note, 6. https://studylib.net/doc/8647594/3d-data-management--controlling-data-volume--velocity–an... Accessed 02 February 2022.

Lohr, S. (2012). The age of Big Data. The New York Times. https://www.nytimes.com/2012/02/12/sunday-review/big-datas-impact-in-the-world.html. Accessed 02 December 2021.

Markoff, J. (2018). Time split to the nanosecond is precisely what wall street wants. The New York Times. https://www.nytimes.com/2018/06/29/technology/computer-networks-speed-nasdaq.html. Accessed 02 February 2021.

Mckinsey GI. (2011). Big data: The next frontier for innovation, competition, and productivity. McKinsey Global Institute. https://www.mckinsey.com/~/media/mckinsey/business%20functions/mckinsey%20digital/our%20insights/big%20data%20the%20next%20frontier%20for%20innovation/mgi_big_data_full_report.pdf. Accessed 02 February 2021.

McKinsey GI. (2016). The age of analytics: Competing in a data-driven world. McKinsey Global Institute. https://www.mckinsey.com/~/media/mckinsey/industries/public%20and%20social%20sector/our%20insights/the%20age%20of%20analytics%20competing%20in%20a%20data%20driven%20world/mgi-the-age-of-analytics-full-report.pdf. Accessed 16 June 2020.

SEC. (2020). Staff report on algorithmic trading in U.S. Capital Markets. Securities and Exchange Commission. https://www.sec.gov/files/Algo_Trading_Report_2020.pdf. Accessed 02 February 2022.

SIFMA. (2019). Electronic trading market structure primer. SIMFA. Securities Industry and Financial Markets Association. https://www.sifma.org/wp-content/uploads/2019/10/SIFMA-Insights-Electronic-Trading-Market-Structure-Primer.pdf. Accessed 02 February 2022.

TCFD. (2017). Final report: Recommendations of the task force on climate-related financial disclosures. https://assets.bbhub.io/company/sites/60/2020/10/FINAL-2017-TCFD-Report-11052018.pdf. Accessed 02 February 2021.

TCFD-PAT. (2021). Measuring portfolio alignment: Technical considerations. https://www.tcfdhub.org/wp-content/uploads/2021/10/PAT_Measuring_Portfolio_Alignment_Technical_Considerations.pdf. Accessed 02 February 2022.

TCFD-PAT, (2020). Measuring portfolio alignment. https://www.tcfdhub.org/wp-content/uploads/2020/10/PAT-Report-20201109-Final.pdf. Accessed 02 February 2022.

TNFD, (2021b). Nature in Scope: A summary of the proposed scope, governance, work plan, communication and resourcing plan of the TNFD. Taskforce on Nature-related Financial Disclosures. https://tnfd.global/wp-content/uploads/2021/07/TNFD-Nature-in-Scope-2.pdf. Accessed 02 February 2022.

Turing, A. (1948). Intelligent machinery: A report. National Physical Laboratory. https://www.npl.co.uk/getattachment/about-us/History/Famous-faces/Alan-Turing/80916595-Intelligent-Machinery.pdf?lang=en-GB. Accessed 02 February 2022.

UNFCCC. (2021). COP 26 and the Glasgow Financial Alliance for Net Zero (GFANZ). https://racetozero.unfccc.int/wp-content/uploads/2021/04/GFANZ.pdf. Accessed 02 February 2022.

8

Sustainable Money Mechanics in Space

You can't depend on your eyes when your imagination is out of focus.

Mark Twain, A Connecticut Yankee in King Arthur's Court, 1889

If we have learned one thing from the history of invention and discovery, it is that, in the long run — and often in the short one — the most daring prophecies seem laughably conservative.

Arthur C. Clarke, The Exploration of Space, 1951

As introduced and discussed in Chapter 4, one of the most pressing challenges we face today is the financing of the Net Zero transition by mid-century. How much will the transition cost and where will the trillions of dollars necessary to reinvent human productivity come from? The public sector? Where debts and credit ratings have already been strained through the extensive borrowing done to counteract the impact of the recent pandemic? Taxes? The Private sector? Banks?

These questions, which have been extensively debated in the media and by several large consulting firms and banks, are of critical importance and relevance to the future of humanity and our survival. Looking at the infrastructural requirements and market transformations needed, the IEA estimates figures to be around \$4 trillion dollars annually by 2030 (IEA, 2021, 18). According to a recent HSBC and BCG report, reinventing supply chains will cost around \$100 trillion (HSBC-BCG, 2021; Murray, 2021). In a more recent report, McKinsey Global Institute estimates that '[c]apital spending on physical assets for energy and land-use systems in the net-zero transition

A. V. Papazian, *The Space Value of Money*,
https://doi.org/10.1057/978-1-137-59489-1_8

between 2021 and 2050 would amount to about $275 trillion, or $9.2 trillion per year on average, an annual increase of as much as $3.5 trillion from today' (McKinsey GI, 2022, viii).

Interestingly, in June 2021, UN Climate Change Executive Secretary Patricia Espinosa expressed her frustration with developed countries who still had to deliver on their promise made in 2010 to allocate $100 billion annually in climate finance to support developing nations. Speaking at the UN Climate Talks Espinosa said: 'We are still talking about this promise, despite greenhouse gas emissions continuing to be at their highest concentration ever; while extreme weather continues to decimate more parts of the world and with greater intensity; and while vulnerable people continue to suffer, continue to lose their livelihoods and their lives' (UNFCC, 2021). If a $100 billion a year is looking hard to materialise, what will become of the trillions necessary to transform the very architecture of our energy systems?

At COP26, the Glasgow Financial Alliance for Net Zero (GFANZ), which includes more than 450 financial institutions from around the world managing $130 trillion of assets, pledged to support the transition (UNFCCC, 2021). The U.N. climate envoy Mark Carney, who assembled the alliance, said that $100 trillion is needed over the next three decades (Jessop & Shalal, 2021). Despite the good intentions and encouraging announcements, these same institutions are still some of the largest investors in fossil fuels, even coal is still an active sector despite pledges (Ainger, 2022; Metcalf & Morales, 2021).

Given our current financial value framework, the history of the industry, and the discussion in earlier chapters, it is safe to say that, as of today, we still do not have a clear and definite funding plan for the transition and we are still searching for an actionable framework that can deploy it effectively.

This chapter discusses a topic that I have often written about, a topic that is central to not only the climate crisis and the transition, but all our evolutionary challenges including, amongst others, outer space exploration.

The topic is money creation in the context of a transformed value framework in finance, exploring the changes necessary to introduce space impact responsibility into our monetary architecture. If investors must earn their returns with a positive space impact, then money creators, whether they are commercial or central banks, must also follow the same principle.

The starting point of the discussion is our current money creation methodology, i.e., debt. The chapter is not a history of money, nor a history of money creation (Davies, 2002; Ferguson, 2008). It is a theoretical discussion focused on addressing the limitations of our current methodology and suggesting an alternative approach that can rectify the shortcomings of a purely debt-based system. The proposed approach introduces new possibilities to help fund our evolutionary challenges with immense risks and very distant returns—two features poorly catered for by our current financial value framework.

I conclude the chapter with a brief discussion of cryptocurrencies given their recent popularity, carbon footprint, and the logic of money creation they reveal.

8.1 Money Creation via Debt

A recent report published by the Bank of England Quarterly Bulletin describes the nature of money simply and accurately:

> There are three main types of money: currency, bank deposits and central bank reserves. Each represents an IOU from one sector of the economy to another. Most money in the modern economy is in the form of bank deposits, which are created by commercial banks themselves (McLeay et al., 2014a, 4).

In the same report, we read: 'broad money is a measure of the total amount of money held by households and companies in the economy. Broad money is made up of bank deposits—which are essentially IOUs from commercial banks to households and companies—and currency—mostly IOUs from the central bank. Of the two types of broad money, bank deposits make up the vast majority—97% of the amount currently in circulation' (McLeay et al., 2014b, 15).

Looking at Fig. 8.1 recreated from the same Bank of England report on money creation in the modern economy, we notice the layers of money creation from the central bank level to commercial banks and consumers.

In a more recent paper the Bank of England further elaborates:

> There are two forms of money most commonly used in the UK. Central bank money is a liability of the central bank. It is available to the public in the form of cash. It is also available to commercial banks in the form of central bank reserves. Private money mainly takes the form of deposits in commercial banks—that is, claims on commercial banks held by the public. This 'commercial bank money' is created when commercial banks make loans to households and companies—referred to as the 'real economy' (Bank of England, 2021a).

In other words, and evidently so, the cycle of debt-based money creation begins at the central bank level and expands through commercial banks. Looking at the Bank of England's Issue Department Balance sheet for the

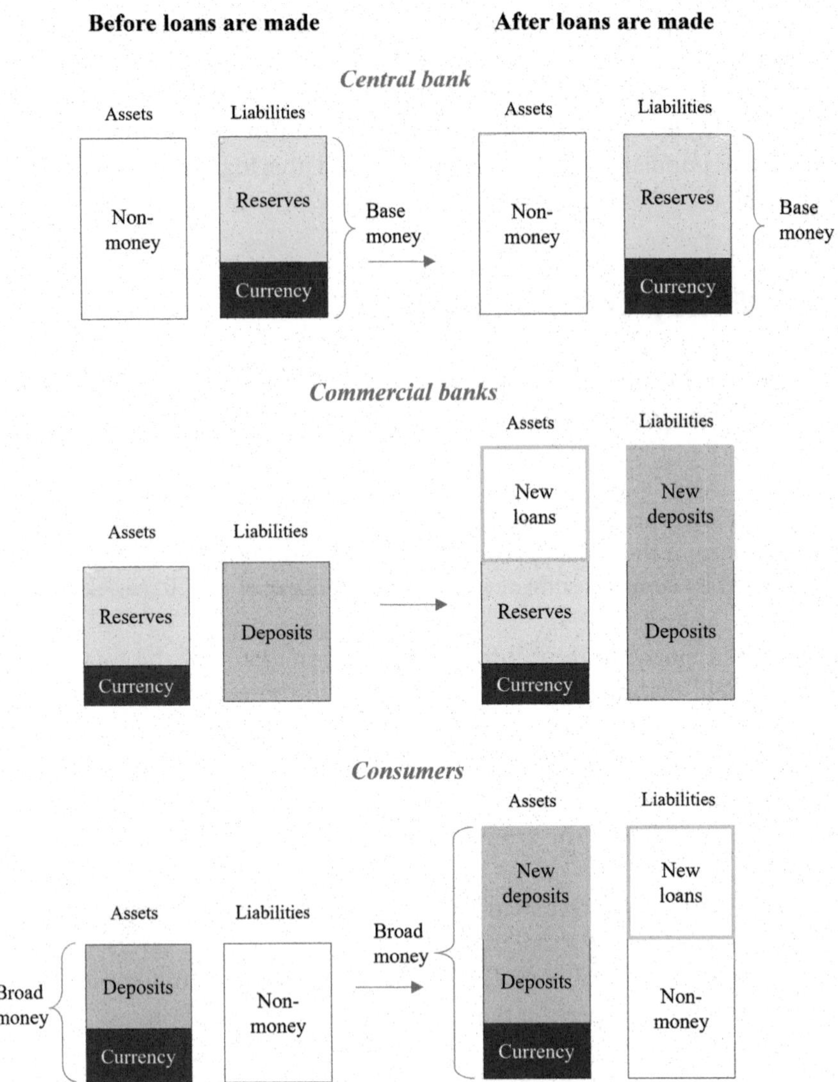

Fig. 8.1 Bank of England money creation process (McLeay et al., 2014b)

period ending 28 of February 2021, we notice that the issued currency (a liability) is backed by government debt securities and 'other securities and assets' (assets).

Looking at the 'other securities and assets' in the notes to the same accounts, we discover the breakdown. Of the £82,599m other assets, £78,524m is a deposit in the Banking Department of the Bank of England. While not all central banks follow this specific methodology, backing currency via a deposit in the other branch of the same bank, for now, it is critical to note that currency at the central bank level is backed by debts and deposits in the UK (Table 8.1).

In 2009, as a response to the financial crisis, HM Treasury authorised the creation of an 'Asset Purchase Facility' (APF) to buy assets and instruments in order to improve liquidity in the market. These purchases were to be financed by the issuance of Treasury Bills, of government debt. In addition, this APF was to be used by the Monetary Policy Committee (MPC) for monetary policy targets. Quoting the 2021 Bank of England Financial statements:

> When the APF is used for monetary policy purposes, purchases of assets are financed by the creation of central bank reserves. The APF transactions are undertaken by a subsidiary company of the Bank of England – the Bank of England Asset Purchase Facility Fund Limited (BEAPFF). The transactions are funded by a loan from the Bank... (Bank of England, 2021b, 117).

In other words, the Bank of England, through a subsidiary company financed by new central bank reserves, i.e., loans, purchases existing high-quality assets from banks, thus increasing reserves, and improving liquidity.

In March 2020, another facility was set up in response to the Coronavirus pandemic, the COVID Corporate Financing Facility (CCFF). Sanctioned by HM Treasury, this facility was designed to support the liquidity of large firms

Table 8.1 BOE Issue Department balance sheet (Bank of England, 2021b)

	2021 £m	2020 £m
Assets		
Securities of, or guaranteed by, the British Government	2,093	2,726
Other securities and assets including those acquired under reverse repurchase agreements	82,599	71,696
Total Assets	**84,692**	**74,422**
Liabilities		
Note Issued		
In Circulation	84,692	74,422
Total Liabilities	**84,692**	**74,422**
Other securities and assets including those acquired under reverse repurchase agreements		
Deposit with Banking Department	78,524	66,552
Reverse repurchase agreements	4,075	5,144
	82,599	**71,696**

in response to COVID disruption. The facility is designed to purchase short-term debt, commercial paper, from large firms whose cash flows may have been affected by the pandemic.

The CCFF is a wholly owned subsidiary of the Bank of England and is operated by the Bank of England, and it is funded through a loan from the Bank of England. However, the Treasury is identified as the 'ultimate risk-owner' of the CCFF and has final say on the eligibility and access of potential corporate issuers (Bank of England, 2021b).

Similarly in the US, the Federal Reserve reacted to both the 2008 and 2020 crises by expanding its balance sheet (FED-BOD, 2020) through new credit facilities and the purchase of debt instruments. In response to both crises the Federal Reserve intervened using all kinds of tools, from reducing the discount rate, the rate at which it lends to banks, to reducing the target of the Federal Funds rate, the rate at which banks lend and borrow to/from each other. The Fed expanded its liquidity enhancing tools through direct purchases of securities from banks, which create new money as reserves. Bernanke, then chairman of the Fed, described the tools used as follows:

> Other than policies tied to current and expected future values of the overnight interest rate, the Federal Reserve has -- and indeed, has been actively using --a range of policy tools to provide direct support to credit markets and thus to the broader economy. As I will elaborate, I find it useful to divide these tools into three groups. Although these sets of tools differ in important respects, they have one aspect in common: They all make use of the asset side of the Federal Reserve's balance sheet. That is, each involves the Fed's authorities to extend credit or purchase securities (Bernanke, 2009).

Bernanke describes the Fed's strategy in 2008 as Credit Easing rather than Quantitative Easing—due simply to the focus on the existing mix of credit conditions. He goes on to describe the difference as follows:

> The Federal Reserve's approach to supporting credit markets is conceptually distinct from quantitative easing (QE).... Our approach -- which could be described as "credit easing" -- resembles quantitative easing in one respect: It involves an expansion of the central bank's balance sheet. However, in a pure QE regime, the focus of policy is the quantity of bank reserves, which are liabilities of the central bank; the composition of loans and securities on the asset side of the central bank's balance sheet is incidental.... In contrast, the Federal Reserve's credit easing approach focuses on the mix of loans and securities that it holds and on how this composition of assets affects credit conditions for households and businesses (Bernanke, 2009)

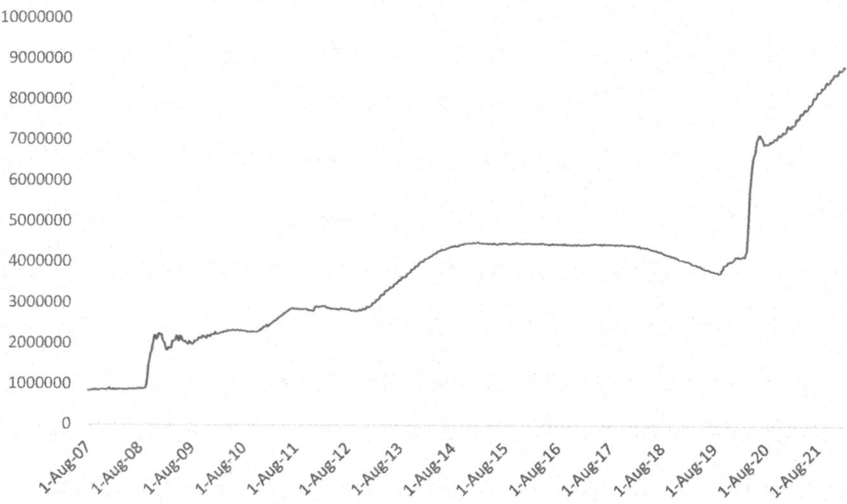

Chart 8.1 Federal reserve balance sheet, assets in $ millions (FED, 2021)

Credit and Quantitative Easing both involve debt-based money creation through the expansion of the Fed's Assets (Curdia & Woodford, 2011). The key difference here is the starting focus of the policy, the Fed's liability side or asset side. Looking at the Federal Reserve Assets in Chart 8.1, we can clearly see the two steep jumps in 2008 and 2020, identifying the injections of liquidity done in response to the 2008 financial crisis and the 2020 Coronavirus pandemic.

After receiving new liquidity from the Federal Reserve, the banks can in their turn extend new loans to businesses and households. Interestingly, as Bernanke describes the pitfalls of this next step in money creation: 'concerns about capital, asset quality, and credit risk continue to limit the willingness of many intermediaries to extend credit, even when liquidity is ample' (Bernanke, 2009).

The evidence provided in this section confirms the debt-based monetary architecture we live in, built on IOUs, at central bank level and commercial bank level.

8.2 The Challenges of Debt-Based Money

An 'I Owe You' as the foundation of money is what debt-based money is. Our current money creation methodology is built on these debt relationships, and the pyramid of IOUs that underpins the world economy has many serious implications.

8.2.1 Calendar Time and Space

Debt, which involves time obligations in terms of scheduled interest and principal repayments, chains everybody involved to calendar time payments. Indeed, whatever the actual shape of the repayment schedule involved, our debt-based money creation methodology chains our entire productive and creative potential to calendar time (Papazian, 2013a).

Governments, government agencies, municipalities, small businesses, households, individuals, corporations, and even banks are all chained to calendar time payments. We are chained to calendar time payments, and directly or indirectly to banks and central banks, the creators of the IOUs through which we have built and structured our economic activities and productivity.

This is true across countries and regions, and though not all manage this money creation architecture at the same level of efficiency and reliability, the debt-based monetary architecture is a global phenomenon.

Note that I do not use the term time, but calendar time. This is important because the nature of time is a debateable subject. Einstein described it as the fourth dimension, revealing the relativity of the experience of time, depending on the observer and the point of view. Physicists have observed the phenomenon of time dilation due to velocity or gravitational potential. Time dilation refers to the difference in elapsed time between two clocks which are affected by their relative velocity and their location close to a large gravitational entity. Carlo Rovelli argues that time is an illusion entirely (Greene, 2004; Rovelli, 2018; Smolin, 2006).

Besides the nature of time, there is also a vast literature in psychology and neuroscience that explores and discusses the perception of time. This has also been applied to money and finance, and is discussed within the context of psychological time, as opposed to calendar or clock time (Allais, 1972, 1974; Blanqué, 2019, 2021). Blanqué (2021, 14) refers to clock time as physical time, the traditional analytical reference of economic and monetary phenomena. He discusses the role and effects of psychological time, 'a time that is perceived or felt', and identifies psychological time as a key determinant. While the discussion is focused on the velocity of money,[1] Blanqué (2021) rightfully reveals the broader limitations of an analytical framework and process built entirely on calendar time.

[1] The effects of psychological time are discussed through duration, memory, and forgetfulness, which define the relative importance of events in shaping expectations and behaviours, irrespective of physical time distances. A unique modality of analysing the impact of the past on present and future expectations (Blanqué 2021).

The time used and applied by banks and central banks in debt transactions is simply calendar time. This is one of the key issues with debt-based money. Linking money creation to calendar time obligations constrains our ability to explore and expand into space timelessly.

The argument proposed here does not downplay the role and importance of calendar time. Calendar time gives structure and direction to our days and our productive activities, and it is a human invention that helps us structure and navigate our productive life on the planet. Calendars and calendar time calculations have been continuously transformed and improved throughout human history.[2] Today, our clocks and calendars rely on a spatial reference point. The conceptual mapping of space and time on Earth makes use of the Prime Meridian, epitomised by the Greenwich laser beam, at 0° longitude. This single point/line in space, on earth, is an important element and structural pillar of the entire world economy (Withers, 2017).

> The Prime Meridian is the line and the point at which the world's longitude is set at 0°. It does not exist in any strict material sense, yet through maps and clocks, the prime meridian governs the life of every human on Earth. (Withers, 2017, 5)

How important can a point and a line be? Without the prime meridian, we would not be able to identify our location, we would not be able to work out time zones, navigate airplanes, land rockets and space capsules, etc. Indeed, we manage our own spacetime through this conceptual projection.[3]

While the prime meridian defines the grid through which we can locate ourselves on Earth and navigate space and time, it does not actually define space, just like calendar time does not define time. Astrophysicists have observed that the universe is continuously expanding, what is known as

---

[2] Indeed, calendar time is a human conception with a diverse history of evolution around the world. The Gregorian calendar, designed by Aloysius Lilius (c. 1510–1576), also known as Luigi Lilio, and introduced by Pope Gregory XIII in October 1582, is a solar calendar and is used by most of the world today and has grown in prominence for the same reasons. Today we have ISO standards that govern date and time notations and calculations. While with religious origins, an adjustment to the Julian calendar, the Gregorian calendar has been adopted by civil authorities around the world. A relatively recent adoption example is Saudi Arabia, which adopted the Gregorian calendar in 2016, but continues to use the Lunar Islamic calendar in parallel (Economist, 2016). Religious and other regional calendars are still used in specific parts of the world, like the Jewish Lunar Calendar, but they are not used as a reference in global banking and commerce today.

[3] The history of the Prime Meridian is fascinating. There was a time in human history when we had more than one 'prime meridian' used by different empires. Recognising the challenge of coordinated communication and trade based on different reference points, the Greenwich Prime Meridian was selected in 1884. See Withers (2017) for a fascinating account of the history behind this central concept that makes our own time and space an intelligible reality. See Malys et al. (2015) for measurement and technical history and measurement differences in the Prime Meridian.

cosmic inflation. The expansion of the universe implies the increase in distance between two entities in the observable universe. Recent findings reveal that the universe is expanding at an expansion rate of 73.3 kms per second per megaparsec, where a megaparsec is 3.26 million light years (Blakeslee et al., 2021).

The point being made here is that the prime meridian, longitudes, latitudes, and calendar time are human inventions, concepts that have been projected onto our reality to allow us to navigate our spacetime, to help us structure our productivity. Our reality, in truth, is far more fluid, far more expansive, and far more malleable.

While using calendar time and the prime meridian to structure and organise our physical movements, location, velocity, communications, etc., is perfectly understandable and useful, linking money creation to such concepts constrains our monetary potential in space by imposing artificially created limits.

Money is a human invention that can and must be continuously fine-tuned and improved. Linking money creation to calendar time based instruments for a species living on a planet in space pauses a serious challenge. It constrains our ability to explore space, to invest in space, to build up space without any concern or consideration of calendar time.

8.2.2 Monetary Gravity and Monetary Hunger

Debt-based money creation, at the central bank level and at the commercial bank level, is achieved through credit and debt instruments that require the repayment of principal and interest to the original source within a specific calendar time window. In other words, it requires a feedback loop to the original starting point. While this is entirely expected and logical, it is also the source of a unique type of artificially created force on our planet. I call this force *monetary gravity*.

This is a human invention, and it acts as a powerful constraint on how far in space an investment can reach before it must return to pay back principal and interest to some bank. Let me demonstrate this through a simple example starting with the speed of light. Think of the distance that light can travel in one month, also known as a light-month, which is a measure of length. It is the distance that light travels in an absolute vacuum in one full month. The speed of light is equal to 299 792 458 m/s, given that we have 30 days in a month, and 86,400 s in each, we can calculate that in one month, light travels 777,062,051,136,000 m, which is equivalent to approximately 777 Tm (1 Terameter = 1,000,000,000,000 m).

The numeric example is simply to illustrate the fact that in one month there is only a certain amount of distance light can travel, before it must return to pay interest to the bank. Assuming a round trip for a physical deposit, the maximum distance light can travel before it needs to return to pay interest is around 389 Tm.

$$Maximum\,Distance_{Light} = Speed\ of\ Light \frac{m}{s} \times Time\ Interval\ in\ s \qquad (8.1)$$

Now, if light, at that speed has a limit as to how far it can travel in a month, how far can a human travel in a month?

$$Maximum\,Distance_{Human} = Speed\ of\ Human \frac{m}{s} \times Time\ Interval\ in\ s$$

Usain Bolt, the Jamaican sprinter, set the world record in 2009 in the 100-m sprint at 9.58 s, giving him a speed of 10.44 m per second. Based on Bolt's speed, the furthest a human can run in one month is 27,060,480 m. Taking the fastest production car as an example, the SSC Tuatara, which is reported to have a speed of 316 miles per hour, or 141.265 m per second, the furthest we can travel within a month is 366,158,880 m. Taking the fastest spacecraft as an example, the Parker Solar Probe (NASA, 2018), which achieved speeds of 153,454 miles per hour or 68,600.076 m per second, the furthest we can travel within a month is 177,811,397,406.72 m. Figure 8.2 gives a visual representation of the limitation on distance travelled.

In other words, when we define the creation of money via instruments that require calendar time commitments, we, in effect, impose a limit on how far

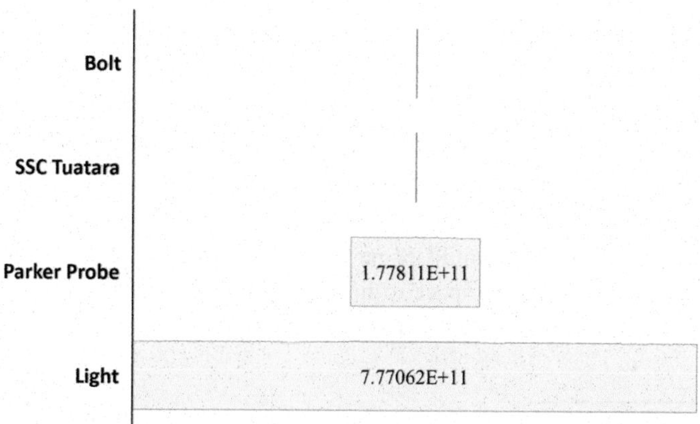

Fig. 8.2 Distance limits due to time imposed limits for one month (*Source* Author)

the financed process can go before it must return to make payments. This obligation to return and pay interest, or obligation to generate the income within a specified calendar time window to make principal and interest payments, is what I call *monetary gravity*.

Debt-based money acts as a leash on our species, chaining us to a self-created calendar, to a self-created system that ultimately chains us to the surface of the planet. Given that calendar time is a human concept, artificially created to manage human activities, linking money creation to an artificially limited concept such as a month or a year, artificially limits the distance we can travel before we need to return to the bank. Naturally, with electronic transfers, we could send in the debt payments without physically returning to the bank, true, but the calendar time limit and pressure to earn the income needed for the payment remain.

The calendar time component in money creation instruments also imposes a limitation on what can be done before payment is due. If there was no time interval imposed, in truth, there would be no limit to the distance travelled, nor there would be any limit on what can be achieved, as the universe is expanding, at 73.3 kms per second per megaparsec, and we have an entire galaxy to explore.

Monetary gravity is far more constraining than actual gravity, where our propulsion technologies have already taken us to the moon and Mars. Monetary gravity chains us to the surface of the planet, spinning around an imaginary calendar, while ignoring the vast landscape of space and the resources within. We are instead forced to deal with limited budgets, credit ratings, and a host of constraints like a debt ceiling in the US (more on this in a later section).

The concept of monetary gravity affects our strategic as well as tactical thinking and actions across many fields and sectors, and in many shapes and forms. One such instance is the concept of 'price of a pound to orbit' which describes the costs of moving one pound from the surface of the planet into LEO, MEO, or GEO orbit.[4] These costs vary depending on the target level of orbit of course. Until recently, the cost to orbit has been estimated to be around $10,000 per pound of payload.

> It costs $10,000 to put a pound of payload in Earth orbit. NASA's goal is to reduce the cost of getting to space to hundreds of dollars per pound within 25 years and tens of dollars per pound within 40 years (Futron, 2002).[5]

---

[4] Low Earth Orbit, Medium Earth Orbit, Geostationary Orbit.

[5] The Futron (2002) report published by the Futron Corporation reports the historical cost per pound to orbit on a historical basis and provides a summary of the key aspects.

NASA's Advanced Space Transportation Program (ASTP) (NASA, 2021) has been working on a number of technological solutions that aim to achieve a continuous decline in cost per pound to orbit. In the commercial space sector, with the recent advancements in EELV (Evolved Expendable Launch Vehicles) technologies pioneered by Space X and others, space transport costs have come down significantly (Chaikin, 2012). Based on Space X's latest figures (Space X, 2022) its Falcon Heavy rocket can deliver to Low Earth Orbit at a pound to orbit cost of around \$640.

On a basic conceptual level, the equation is measuring the cost of launch and the load deployed through the launch.

$$\textit{Cost per Pound to Orbit} = \frac{\textit{Cost of Launch}(\$)}{\textit{Load Deployed}(lb)} \tag{8.2}$$

The link of this discussion to the nature of money is through the focus on the \$/lb to orbit equation. Moving an x mass across space is being linked to y dollars, and y dollars are linked to z budgets and w debts. The monetary gravity reveals itself through the very existence of the concern.

Instead of the \$/lb to orbit equation, we should be focused on the most efficient and fastest technology to transport maximum payloads in the greenest and safest way possible, in the highest frequency possible. Indeed, we should probably pay more attention to the carbon footprint of the rocket equation rather than the cost. This subject has attracted significant attention recently.

$$\textit{Emission per Pound to Orbit} = \frac{\textit{Emissions of Launch}(\$)}{\textit{Load Deployed}(lb)} \tag{8.3}$$

Based on the discussion and main propositions of the space value framework, we should, in fact, consider the overall space impact per launch per pound to orbit. Moreover, we should also consider the space impact of the payload itself. We can use the Net Space Value of the payload and the launch for this purpose, allowing us to include a number of different aspects of impact through the payload as well as the launch itself. Naturally, we should also pay special attention to negative impacts.

$$\textit{Space Impact per Pound to Orbit} = \frac{\textit{NSV of Launch}(\$) + \textit{NSV of Payload}(\$)}{\textit{Load Deployed}(lb)} \tag{8.4}$$

When we link the strategic thinking that underpins our innovative technologies to a logic of cost reduction, driven by the constraints on public budgets and lack of immediate markets in the present, we are indeed serving monetary gravity. When, in truth, money should never have been relevant to the logic of the invention, as it is an artificial optimisation target introduced thanks to a debt-based time-linked monetary architecture, on the one hand, and commercial market pressures on the other.

The concept of monetary gravity is only half the story. Debt-based money also creates a *monetary hunger* in the system. Given that money is continuously being created through debts at the central bank and commercial bank levels, at any point in time and in any economy, there is always a large group of individuals, households, businesses, corporations, municipalities, governments, and banks that are chasing deposits and cash to earn enough income to pay back principal and interest payments on their loans. This calendar time linked chase and competition for notes and deposits to pay back debts is what I refer to as *monetary hunger*.

Monetary hunger is a significant driver of growth in our current economic system. All borrowers must achieve and earn enough to pay interest and principal, and they must compete for the available money to earn enough to make those payments. The need to fulfil contractual debt obligations, the legal threat of default, and potential loss of collateralised assets or credit ratings act as a significant stimulant for individuals, households, corporates, governments, and banks.

Although rolling over debt is a mechanism through which some of the above-mentioned entities can reduce the pressures of time-linked debt obligations, this serves to kick the can down the road, but does not ultimately change the logic of the system and the instruments used. Indeed, this monetary hunger to fulfil debt obligations in the system often triggers a competitive and careless drive for growth.

Recently, the Institute of International Finance (IIF) published a report (IIF, 2021) where global debt levels were reported to be close to $300 trillion. A Reuters news discussing the report states 'Total debt levels, which include government, household and corporate and bank debt, rose $4.8 trillion to $296 trillion at the end of June, after a slight decline in the first quarter, to stand $36 trillion above pre-pandemic levels' (Ranasinghe, 2021).

The more we create money via debt-based instruments, the more we chain ourselves to calendar time, and speed up the race to fulfil capital and interest payments. Interestingly, this is true independently of the levels of capital accumulation in the system. Given the debts and the interests and the link to calendar time, both monetary gravity and monetary hunger remain in the system, chaining us to calendar time, to the surface of the planet, and triggering the financial and business strategies that consume the planet.

8.2.3 Innovation, Risk, and Control

When we create money via debt, via calendar time dependent instruments, we are indeed constraining the deployment and return of the invented money through the calendar time obligations attached to the instrument. After all, missing those payments can lead to default and bankruptcy, which implies that debt-based money must be deployed and repaid in a way that secures its calendar time obligations.

When using debt instruments to create money, we are always faced with a mismatch between the rigid calendar that keeps moving forward at a predefined and uniform rate, and true innovation. Innovative breakthroughs happen because of intense thinking, testing, brainstorming, experimenting, and often, significant revisions. All the above can be crammed within a calendar time window but may or may not deliver results within it. Indeed, there is a vast literature that explores the sensitivity of innovation to different types of financing.

The point here is to recognise that a calendar time linked money creation process puts the onus of managing this mismatch between calendar time and innovation time squarely on borrowers. While the inventors of money, i.e., banks and central banks, enjoy a well-priced risk secured through appropriate safeguards and collaterals.

To clarify the relationship to risk and control, let's take the example of an equity investment or instrument. An equity investment, i.e., buying shares in a company, unlike debt, does not create such time-based obligations that can lead to bankruptcy or default. Even when promises are made through milestones, an equity investment cannot lead to the forced foreclosure or repossession of the assets created through the investment. While debt contracts and bank loans can.

Interestingly though, equity investments do not involve the creation of new money. They involve the transfer of existing deposits from account to account, from investor to company, from investor to investor, etc., but do not involve the creation of new deposits or banknotes.

The type and level of risk and control involved in a debt transaction are very different from an equity transaction. As such, banks, central and commercial, would have a very different relationship to the assets when they are created via a debt transaction vs when they are created through an equity transaction. They would also have a very different risk exposure.[6]

Equity investments, especially public ones, do impose time-based reporting and disclosures, but they do not chain the investment and the process to calendar time the way a debt transaction does. Moreover, they share the risks of the investment very differently. In equity transactions, risk is shared with the project or company for a return that will also be shared. In the case of debt, secure or unsecured, given laws of bankruptcy and liquidation in most of the world, creditors are paid before shareholders, and thus lenders come before shareholders. Similarly, interests on loans are paid before dividends on shares.

Furthermore, because debts are often given on a collateralised basis, individuals and corporations must often provide assets as collateral in order to be able to borrow. Mortgages by the trillions are backed by the house, apartment, or real estate being bought or developed. Personal loans, aside from credit cards and student loans, are also often backed by assets, and the same applies to business loans. This is why the industry has a commonly used concept like the Loan to Value ratio, which is used by the lender to decide what percentage of the asset's value they are willing to lend. A 75% LTV ratio implies that the bank or lender will lend 75% of the value of the asset being provided as collateral. Usually, the LTV ratio depends on the 5 Cs of credit or creditworthiness, i.e., character, capacity, capital, collateral, and conditions.

The point being made here is that debt-based money creation has a unique risk and control equation vis-à-vis the created assets and affects innovation potential and timelines.

To conclude, debt-based time-linked money creation chains our money creation methodology to a limited concept, i.e., calendar time, and by default creates what I have described as monetary gravity and hunger within the system. Furthermore, it has a restrictive impact on innovation and establishes an unbalanced risk/control equation in the process.

8.3 Value Easing: Sustainable Money Creation

In this section, I propose and discuss a new logic of money creation built on space value creation, which respects the space value of money principle,

[6] I would like to note here that Islamic finance, in principle at least, requires that capital earns a return rather than interest. As such, in principle, Islamic instruments require that investments involve productive activity, sharing risks and profits. In an article I wrote in 2011, titled, Islamic Money Creation, I proposed that Islamic finance principles can help develop or support a new approach to monetisation, involving equity-like features based on risk sharing (Papazian, 2011b).

and has a very different time, risk, and asset profile. Such a change in money mechanics can be an empowering tool when addressing evolutionary challenges like the transition to Net Zero, the eradication of poverty, and the establishment of a permanent human habitat on the moon.

The changes proposed here are built on the following premises: (1) we can change and improve the methodology behind money creation, (2) we can create money through instruments that do not chain us to calendar time, (3) we can create money through instruments that support our evolutionary ambitions without artificially created monetary gravity, (4) we can create money through instruments that do not create monetary hunger in the system, and (5) we can create money through instruments that share risks and assets while supporting innovation.

While the change I propose is highly significant from a systemic point of view and can open yet inconceivable opportunities to fund our evolutionary ambitions, from an operational, institutional, accounting, and technological perspective, what I propose is a very minor and functional change.

We can introduce a new type of financial instrument for the purpose of money creation. After all, if the Bank of England can create and back banknotes by a deposit in the banking department of the Bank of England, if the Bank of England can create new money by loaning to its own wholly owned subsidiary, if the Federal Reserve can create new money by buying toxic Collateralised Debt Obligations and Mortgage-Backed Securities or by buying commercial paper, there is no reason why they cannot back or create new money through an alternative equity-like instrument that shares risks, shares the ownership of the assets created through the instrument, has a tangible and inspiring positive space impact, and helps resolve our evolutionary challenges.

Naturally, I am using the Bank of England and the Federal Reserve as examples here, this applies to all central banks. To achieve such a change, we must accept that we can diversify the instruments used for money creation, and through such an operational transaction, improve an entire system. Indeed, we have always created new financial instruments, some useful and some not so useful, and there is no reason to exclude innovative financial engineering from transforming our money creation methodology.

One more point to clarify before digging deeper into the new alternative instrument that I propose. I am not suggesting here that we should 'print' our way out of the challenges we face, although that is exactly what the Fed and the Bank of England and the European Central Bank and others have been doing lately. The suggestion here is to inject strategically through non-debt instruments into key sectors that can change the course of human history and help us address challenges such as climate change. I call this Value Easing, in contrast and in parallel to credit and quantitative easing, and I define it as follows:

Value Easing: The transactional process undertaken by a central bank that consists in purchasing non-debt no-maturity equity-like high space impact value creating instruments from qualified government agencies and/or public–private partnerships (PPP) with relevant Treasury sponsorship that increases the central bank's balance sheet and injects new liquidity outside the banking system.

In truth, injecting new money through value easing defined above, and described in the following sections, will also reduce the currently experienced inflationary pressures. This is so because the injected new money is immediately allocated to new value creation and real investment where they are most needed, which is not necessarily the case with credit and quantitative easing.

8.3.1 From Debt to Space Value Creation

The logic of money creation in a debt instrument is the commitment to repay the invented deposit or money. The trigger of money creation is the repayment commitment and schedule. In the case of banks, the new deposits are created as soon as the client's signature is on the loan contract which becomes a binding document through which they secure their payments. The loan document becomes an asset on the bank's balance sheet, and the created deposit becomes a liability. When Central Banks create new money through lending or the purchase of securities, the same logic applies. They are adding debt instruments to their asset side and currency or reserves on their liabilities side. The commitment and schedule of interest payments or coupons and principal at maturity are central to the transaction.

Indeed, we have an entire credit rating industry that exists for the purpose of gauging repayment probability, capability, and risks. This is a reality in the public and private spheres, for individuals, businesses, governments, and banks.

In the suggested change that I propose here, the new alternative instrument introduces a new logic, a new trigger, a new type of commitment. The proposed instrument changes the trigger of money creation from *commitment to pay back* to *commitment to create responsibly and pay a return when possible*. Such a change would imply that the trigger for money creation is not the repayment of the invented money, but the creation and sharing of space value through the invented money. The key difference here is the sharing of risks and the ownership of the assets created, or the relationship to and with the assets being purchased or created through money creation.

I have proposed elsewhere and on numerous occasions that we can introduce equity-like non-debt high space value impact instruments for money creation purposes, both as a monetary policy tool as well as a channel through which real productive activity can be initiated and directed towards the resolution of our key evolutionary challenges. I have called them Public Capitalisation Notes (PCN), instruments that can improve a system by redirecting the flow of money creation while optimising space impact (2013a).

As described in Fig. 8.3 the result of injecting money through debt instruments vs PCNs is ultimately the same, increased bank reserves. The route of injection changes and the relationship to the assets created changes. Via PCNs, the injection creates income and expenditure first through direct investments into evolutionary projects. In the case of debt instruments, the current method, the injection starts as bank reserves, which end up as expenditure and income through the new loans given out by banks.

Public Capitalisation Notes are conceived as financial instruments that would be issued by a relevant government agency and sponsored by the Treasury. A PCN can be just as valid as a Mortgage-Backed Security (MBS) or a AAA corporate bond for a central bank to use for their injection of new money. Nothing really prevents us from channelling newly invented money into a transition administration agency, rather than the big banks, except a debt-based logic of money creation. Indeed, we could allocate newly

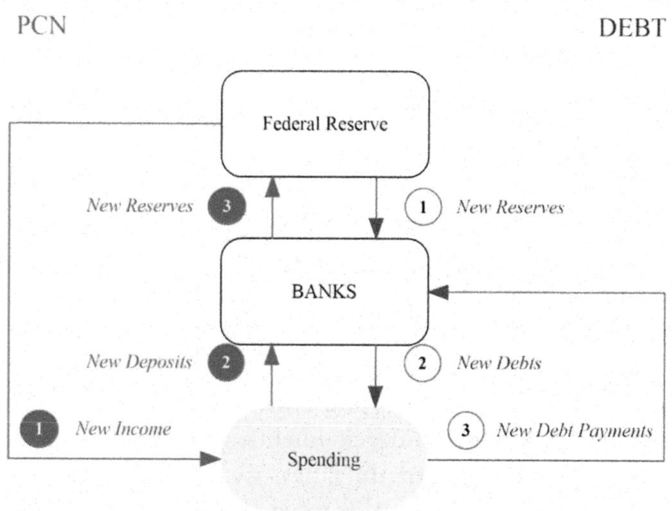

Fig. 8.3 PCN vs DEBT injection: new income vs new loans (*Source* Author)

invented money through relevant structures and administrative agencies to the International Energy Agency to oversee the energy transition we need.

8.3.2 Climate PCN

Figure 8.4 describes the broad architecture of a Climate PCN. The instrument is issued by an administrative agency identified as the ideal manager/host of the injection, it is sponsored by the relevant treasury(ies) and purchased directly by the central bank(s). It is aimed at raising funds to address clear evolutionary targets and challenges. For a Climate PCN, the purpose would be to invest and develop the technologies and infrastructures needed for a transition to a Net Zero economy. This can be done on national and international levels, involving multiple agencies, treasuries, and central banks.

The identification of priorities is set at the administration agency level but built and developed in consultation with industry. Thus, the operational, informational, technological, and financial architecture that is required to implement a Climate PCN is significant and extends far beyond one agency and its capabilities.

PCNs must be managed as economy wide investment tools involving the public and private sectors. They should not be treated or understood as government bonds or corporate bonds where the proceeds are managed according to public and private budgets and agendas. PCNs must be conceived and deployed as strategic investments aimed at achieving thrust in our efforts to address key evolutionary challenges and/or ambitions.

8.3.3 NASA PCN

The current funding architecture behind our outer space exploration imperative poses a challenge—we have relatively meagre debt-based government budgets supporting governmental space agencies (and recently space forces in the US), several large corporates, and billionaire funded companies who feed on those budgets as they try to develop relevant technologies and push for new revenue sources through commercial space applications and tourism. While private money is being more and more attracted to outer space exploration and related technological developments, given the size of the funding needs, the risks, and the return timelines, we are far from providing the necessary investments to this critical frontier (Papazian, 2013a, b, c).

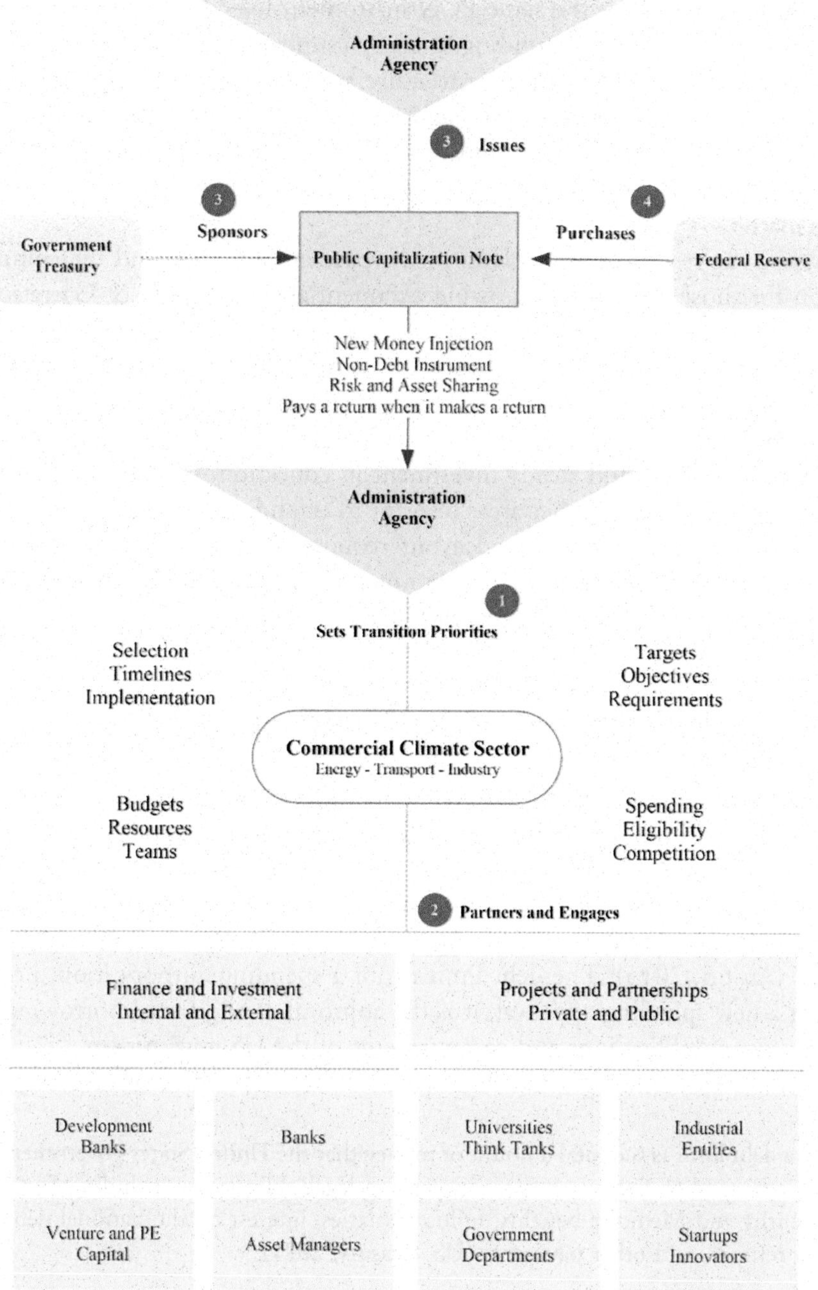

Fig. 8.4 Climate PCN architecture: to finance massive transition to net zero (*Source* Author)

Figure 8.5 describes the same PCN instrument logic but applied to outer space exploration and the funding of an investment drive that aims to build the space technologies and infrastructure we need to clean the debris in orbit, build habitats on the moon, and explore our universe responsibly. The example uses NASA as the administration agency, but it can also be the UK Space Agency, or the European Space Agency, or any other space agency for that matter.

While the resources available in outer space are immense and the competition for those resources is growing exponentially (Goswami & Garretson, 2020),[7] we are yet to match the opportunity with the investment it deserves. Given the return and risk features of outer space exploration, current financial models, indeed, current money creation methodologies, cannot properly cope with space technology financing. Outer space exploration requires significant and steady investment in education, technology, science, industry, and a host of other areas relevant to extending our reach. Financing this high-risk distant-return endeavour requires a sustainable, broad-based, welfare-enhancing investment programme that can be achieved through an appropriately designed and executed PCN.

8.4 From Debt Ceiling to Wealth Floor

Since the 1960s the US Congress has raised the debt limit or ceiling of the US government 78 times (US Treasury, 2021). The concept of an aggregate debt limit or ceiling was introduced by the US Congress much earlier through the Second Liberty Bond Act. Previously, procedural limitations to debt issuance always existed of course, but the aggregate limit or ceiling was introduced in 1917 (Austin, 2015). The debt limit is not a spending authorisation, i.e., it is not a new spending approval, it is the approval of additional borrowing to meet existing obligations and commitments of the US government.

The US Department of the Treasury describes it as follows (Table 8.2):

> The debt limit is the total amount of money that the United States government is authorized to borrow to meet its existing legal obligations, including Social Security and Medicare benefits, military salaries, interest on the national debt, tax refunds, and other payments (US Treasury, 2021).

[7] Goswami and Garretson (2020) in their book, Scramble for the Skies: The Great Power Competition to Control the Resources of Outer Space, provide a detailed review and analysis of the vast outer space resources and the opportunities and challenges that they represent for humanity and the many states in competition to master the technologies necessary to acquire them.

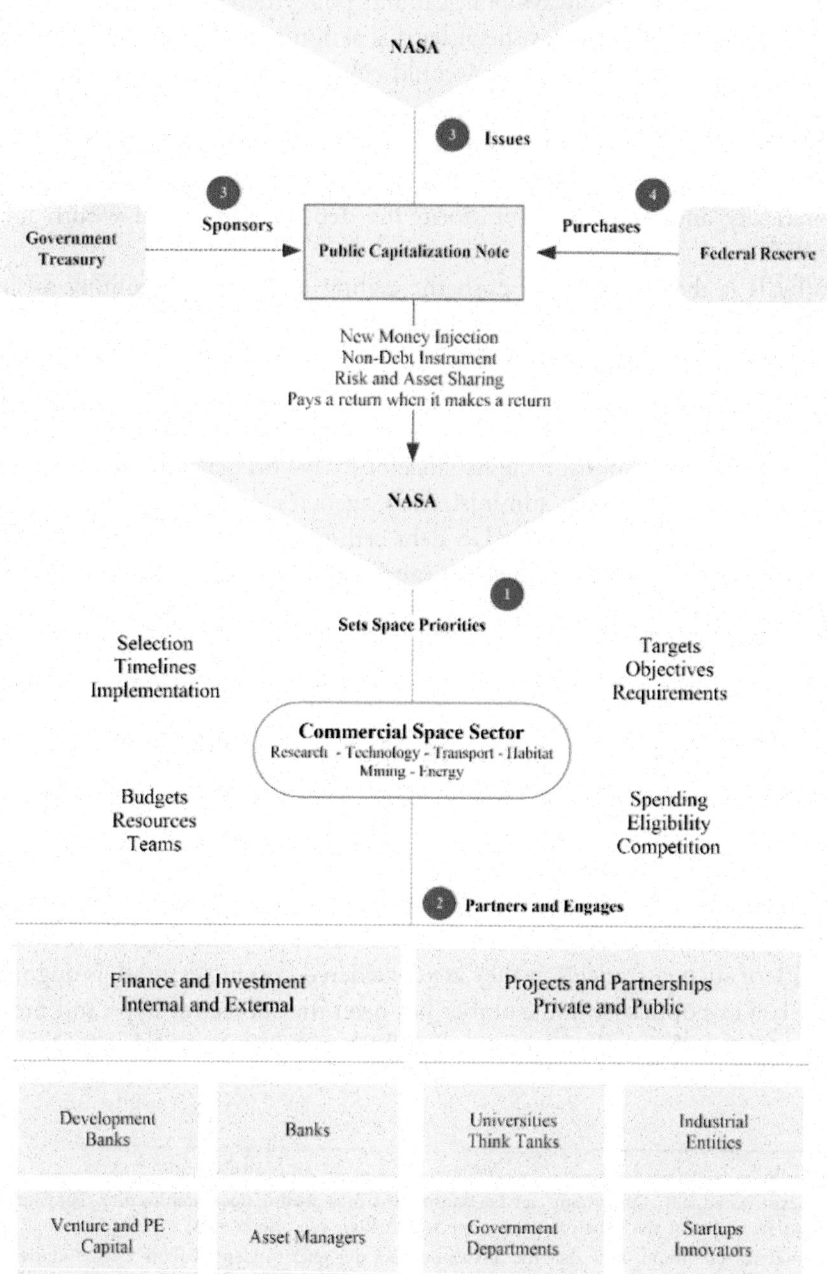

Fig. 8.5 NASA PCN: to finance massive space expansion effort[10] (*Source* Author)

Even though there are always political and policy debates around a request to raise the debt limit, across the aisle, this action is recognised as a systemic necessity, as withholding approval could cause a default of the US government. This debt ceiling is a debt ceiling rather than a wealth floor specifically because of the debt-based monetary architecture we are bound by. If the US government could raise money via non-debt instruments, like PCNs, it could theoretically and practically transform the debt ceiling into a wealth floor (Fig. 8.6).

Why is it that we have to push the ceiling up instead of simply raising the floor? The entire exercise is counterintuitive and an obstacle that exists simply because of a debt-based architecture. If money creation were to be a space value creation exercise rather than debt issuance and purchase exercise, the shift from debt ceiling to wealth floor would become a natural one. If the US Treasury would sponsor the issuance of PCNs on behalf of NASA or the Space Force, or other key administration agencies, and the Federal Reserve would purchase the PCNs, the US debt ceiling can soon become economic history. Congress would not need to raise the debt ceiling and could instead authorise the issuance of new space value creating instruments like PCNs, raising the wealth floor.[8]

8.5 'Money' Creation via Mathematical Guesswork

Before concluding this chapter, I must discuss the cryptocurrency phenomenon. Over the last decade or so we have seen and witnessed the growth of cryptocurrencies like Bitcoin.[9] The debate regarding the reliability and risk of such currencies, if they are considered currencies at all, is ongoing. Their rise in popularity and number has been the subject of hype and media debates about their value, nature, reliability, and use. As of July 2022, there

[8] Naturally, as we start issuing the first PCNs, government debts and Public Capitalisation Notes will coexist. Gradually, the system can be transitioned to a wealth floor architecture. For systemic preservation purposes, the debt ceiling and the wealth floor can also coexist.

[9] To avoid any confusion, note that the discussion does not refer to Central Bank Digital Currencies (CBDC). The Federal Reserve describes CBDCs as '"Central bank money" refers to money that is a liability of the central bank. In the United States, there are currently two types of central bank money: physical currency issued by the Federal Reserve and digital balances held by commercial banks at the Federal Reserve. While Americans have long held money predominantly in digital form—for example in bank accounts, payment apps or through online transactions—a CBDC would differ from existing digital money available to the general public because a CBDC would be a liability of the Federal Reserve, not of a commercial bank' (FED, 2022).

Table 8.2 Summary of us treasury securities outstanding, November 30, 2021, $mn (Treasury Direct, 2021)

Summary of treasury securities outstanding, November 30, 2021			
	Debt Held By the Public	**Intragovernmental Holdings**	**Totals**
Marketable:			
Bills	3,784,677	1,420	3,786,096
Notes	12,845,781	8,780	12,854,561
Bonds	3,426,188	7,337	3,433,525
Treasury Inflation-Protected Securities	1,694,646	626	1,695,272
Floating Rate Notes	576,028	70	576,099
Federal Financing Bank	0	6,053	6,053
Total Marketable	22,327,321	24,286	22,351,607
Nonmarketable:			
Domestic Series	28,592	0	28,592
Foreign Series	264	0	264
State and Local Government Series	113,852	0	113,852
US Savings Securities	144,171	0	144,171
Government Account Series	28,462	6,237,789	6,266,251
Other	3,250	0	3,250
Total Nonmarketable	318,591	6,237,789	6,556,380
Total Public Debt Outstanding	**22,645,912**	**6,262,075**	**28,907,987**
Statutory debt limit, November 30, 2021			
	Debt Held by Public	**Intragovernmental Holdings**	**Totals**
Debt Subject to limit:			
Total Public Debt Outstanding	22,645,912	6,262,075	28,907,987
Less Debt Not Subject to Limit:			
Other Debt	478	0	478
Unamortised Discount	5358	14,661	20,018
Federal Financing Bank	0	6,053	6,053
Plus Other Debt Subject to Limit:			
Guaranteed Debt of Government Agencies	*	0	

(continued)

[10] I have provided an early version of this graphical flow diagram in a Harvard Business School consulting project interview done for the Defense Innovation Unit in the US.

Table 8.2 (continued)

Summary of treasury securities outstanding, November 30, 2021			
Total Public Debt Subject to Limit	22,640,077	6,241,361	28,881,438
Statutory Debt Limit			28,881,463
Balance of Statutory Debt Limit			25

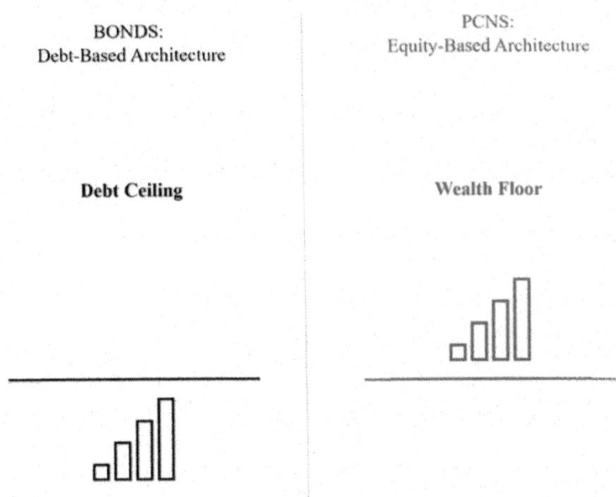

Fig. 8.6 Debt ceiling vs wealth floor (*Source* Author)

were 9931 cryptocurrencies listed on coinmarketcap.com (CoinMarketCap, 2021).

The birth of Bitcoin in 2009 and the subsequent growth in cryptocurrencies have often been ascribed to the disappointment and frustration people felt with banks or debt-based money. Indeed, it is not surprising that their rise has followed the 2008 financial crisis. Whatever the reasons behind their recent popularity, cryptocurrencies have a number of key features that should be discussed. This is so specially in the context of sustainable finance.

The first question concerns their nature, are cryptocurrencies actual currencies or assets, or neither? The debate is ongoing, and the views are diverse and often diametrically opposed. Given their very volatile prices, cryptocurrencies are not ideal methods of payment, and they are quite expensive in relative terms. This is why the Bank of England refers to them as cryptoassets rather than currencies.

> Put it this way, you wouldn't use cryptocurrency to pay for your food shop. In the UK, no major high street shop accepts cryptocurrency as payment. It's

> generally slower and more expensive to pay with cryptocurrency than a recognised currency like sterling. Development is underway to make cryptocurrency easier to use, but for now it isn't very 'money-like'. This is why central banks now refer to them as "cryptoassets" instead of "cryptocurrencies".
>
> Today cryptocurrencies are generally held as investments by people who expect their value to rise (Bank of England 2021c).

The last sentence of this quote hints at an important feature of cryptoassets. They do not have any intrinsic value, and they do not involve any return accruing to holders. Prasad (2021) refers to this by referencing the 'greater fool' theory. He writes: 'The valuations of meme currencies seem to be based entirely on the "greater fool" theory—all you need to do to profit from your investment is to find an even greater fool willing to pay a higher price than you paid for the digital coins'. An alternative interpretation is that the attractiveness of Bitcoin and other cryptocurrencies is in their lack of transparency, as 'dark money' (Economist, 2022).

Another critical aspect of cryptocurrencies is the energy consumption attached to the mining process that generates the coins. The Cambridge Bitcoin Energy Consumption Index (CBECI) estimates that the yearly average annualised electricity consumption of the Bitcoin Network is around 135.3 TWh per year, which is higher than the yearly consumption of Sweden at 123.249 TWh per year (CCAF, 2022).[11]

The below chart provides the monthly (right axis) and cumulative (left axis) electricity consumption of the Bitcoin Network. Chart 8.2 reveals an increasing monthly and cumulative consumption.

Besides the electricity consumption and the resulting high carbon footprint of Bitcoin, a recent study also reveals the electronic waste generated due to the consumption of electronic hardware used in mining the coins. The authors estimate that Bitcoin's e-waste is 30.7 metric kilotons per annum as of May 2021 (De Vries & Stoll, 2021).

All the issues discussed above are relevant to understanding Bitcoin and other cryptocurrencies. The most critical, however, is the very logic of their creation. While central bank and commercial bank money are created via debt transactions, cryptocoins are created after a mining process through powerful computers performing specific mining operations which consist in solving complex mathematical puzzles.

Bitcoin.org describes Bitcoin as money following the below logic:

[11] For a relative understanding of the numbers, note that the two highest consuming countries, China and USA, use, respectively, 6875.09 TWh and 3843.83 TWh per year (CCAF 2022).

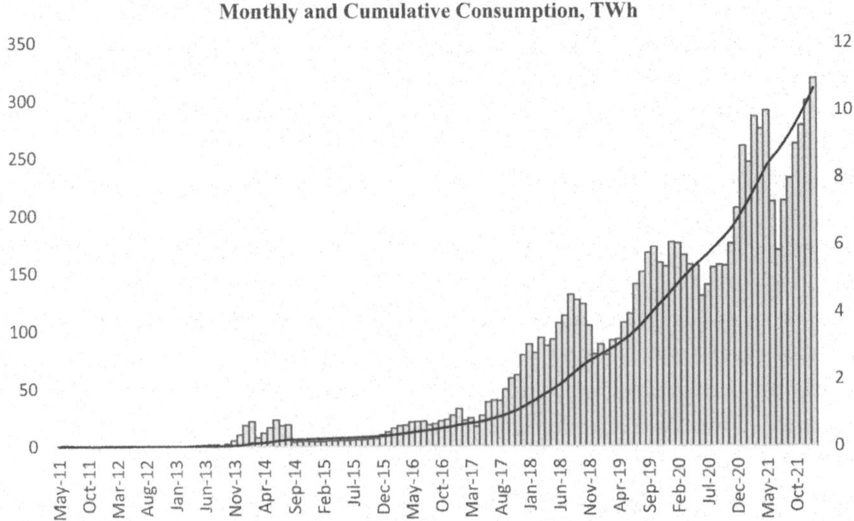

Chart 8.2 Total bitcoin electricity consumption, monthly and cumulative (CCAF, 2022)

> Bitcoins have value because they are useful as a form of money. Bitcoin has the characteristics of money (durability, portability, fungibility, scarcity, divisibility, and recognizability) based on the properties of mathematics rather than relying on physical properties (like gold and silver) or trust in central authorities (like fiat currencies). In short, Bitcoin is backed by mathematics (Bitcoin, 2021a).

While the statement 'Bitcoin is backed by mathematics' is at best misleading, and the above omits some of the key functions of money, let's look into how they are created. Once again, Bitcoin.org describes the process:

> New bitcoins are generated by a competitive and decentralized process called "mining". This process involves that individuals are rewarded by the network for their services. Bitcoin miners are processing transactions and securing the network using specialized hardware and are collecting new bitcoins in exchange (Bitcoin, 2021b)

Bitcoin mining, which is the process through which a new Bitcoin is created and awarded, is the process where we can find the logic of creation. Bitcoin.org describes the process as follows:

> Anybody can become a Bitcoin miner by running software with specialized hardware. Mining software listens for transactions broadcast through the peer-to-peer network and performs appropriate tasks to process and confirm these

> transactions. Bitcoin miners perform this work because they can *earn transaction fees paid by users for faster transaction processing, and newly created bitcoins issued into existence according to a fixed formula.*
>
> For new transactions to be confirmed, they need to be included in a block along with a mathematical proof of work. *Such proofs are very hard to generate because there is no way to create them other than by trying billions of calculations per second*. This requires miners to perform these calculations before their blocks are accepted by the network and before they are rewarded. As more people start to mine, the difficulty of finding valid blocks is automatically increased by the network to ensure that the average time to find a block remains equal to 10 minutes. As a result, mining is a very competitive business where no individual miner can control what is included in the block chain (Bitcoin, 2021a).[12]

The underlying process through/for which Bitcoins are created/awarded involves and is dependent on '*trying billions of calculations per second*'. In other words, to put it in the most benign way possible, minors get rewarded with bitcoin for mathematical guesswork. Basically, from a debt logic, we have now moved to a more preposterous logic of money creation, guesswork.

Besides the shortcomings that are due to a massive carbon and electronic waste footprint, volatility and lack of intrinsic value, the most critical of all the challenges of crypto as currency is the underlying 'work' and logic of their creation. If we are to improve the logic of money creation beyond debt, mathematical guesswork cannot possible be that improvement.

8.6 Conclusion

Once we have transformed our financial value framework and introduced space and our responsibility for impact into our value paradigm, the most immediate next step is to require that along with public and private investors, money creators also adopt the principle and ensure that the process of money creation avoids negative impacts and optimises positive impacts.

Diving deeper into our current monetary architecture, where money is created through public and private debts, at central bank level and at commercial bank level, I identified and discussed a number of key limitations of debt-based money, i.e., features that limit and undermine human productivity in space. Indeed, debt-based money acts like a calendar time

[12] Emphasis added.

linked leash that keeps us chained to the surface of the planet. I also introduced the concepts of monetary gravity and monetary hunger as the limiting forces born from a debt-based money creation methodology.

The chapter proposed a simple transactional approach to transforming debt-based money creation. The idea is to start with the introduction of a non-debt equity-like space value creating instrument that can be used for the purpose. I called this Value Easing, in line with Quantitative Easing and Credit Easing—two debt-based mechanisms for new money injection. I proposed Public Capitalisation Notes (PCN) as one such possible instrument that could be used to finance key evolutionary investments like the transition to Net Zero, the eradication of poverty, and outer space exploration. Naturally, all our environmental challenges can be added on the list.

If the Bank of England, the Federal Reserve, and the European Central Bank can invent and inject money through debt instruments, even toxic instruments when needed, they should be able to do the same through instruments that have high positive space impact that aim to address our evolutionary challenges.

The space value of money principle and the associated metrics can open the door to new funding structures that can facilitate the financing we need to address the many evolutionary challenges we face. They can help us devise the instruments necessary to create and deploy the trillions we need to reduce emissions, to transform our energy systems, to clean the plastic in our oceans, the radioactive waste buried in landfills and in sub-seabed disposal sites, the polluted rivers, and the millions of debris orbiting the Earth. The space value framework can also help raise and deploy the many more trillions needed to eradicate poverty, and fund humanity's next big leap into the solar system.

References

Ainger, J. (2022). Coal is still raising trillions of dollars despite green shift. BNN Bloomberg. https://www-bnnbloomberg-ca.cdn.ampproject.org/c/s/www.bnnbloomberg.ca/coal-is-still-raising-trillions-of-dollars-despite-green-shift-1.1723066.amp.html. Accessed 15 February 2022.

Allais, M. (1972). Forgetfulness and interest. *Journal of Money, Credit and Banking*, *4*(1), 40–73. https://doi.org/10.2307/1991402. Accessed 02 March 2022.

Allais, M. (1974). The psychological rate of interest. *Journal of Money, Credit and Banking*, *6*(3), 285–331. https://doi.org/10.2307/1991172. Accessed 02 March 2022.

Austin, A. (2015). The debt limit: History and recent increases. Congressional Research Service. https://sgp.fas.org/crs/misc/RL31967.pdf. Accessed 02 February 2020.

Bank of England. (2021a). New forms of digital money. Bank of England Discussion Papers. Bank of England, London. https://www.bankofengland.co.uk/paper/2021a/new-forms-of-digital-money. Accessed 02 February 2022.

Bank of England. (2021b). Bank of England annual report and accounts 20/21. Bank of England. https://www.bankofengland.co.uk/-/media/boe/files/annual-report/2021b/boe-2021b.pdf?la=en&hash=965204F6565CB8CAD29A86E595CB7F02E8A54E07#page=97. Accessed 02 February 2022.

Bank of England. (2021c). What are cryptoassets (cryptocurrencies)? Bank of England, England. https://www.bankofengland.co.uk/knowledgebank/what-are-cryptocurrencies. Accessed 02 February 2022.

Bernanke, S. B. (2009). The crisis and the policy response. Federal Reserve Board of Directors. Speech at London School of Economics. https://www.federalreserve.gov/newsevents/speech/bernanke20090113a.htm. Accessed 02 February 2021.

BIS. (2003). The role of central bank money in payment systems. Bank for International Settlements. https://www.bis.org/cpmi/publ/d55.pdf. Accessed 02 February 2021.

Bitcoin. (2021a). Why do Bitcoins have value. Bitcoin.org. https://bitcoin.org/en/faq#why-do-bitcoins-have-value. Accessed 12 December 2021a.

Bitcoin. (2021b). How are bitcoins created? Bitcoin.org. https://bitcoin.org/en/faq#how-are-bitcoins-created. Accessed 12 December 2021b.

Bitcoin. (2021c). How does Bitcoin mining work? Bitcoin.org. https://bitcoin.org/en/faq#how-does-bitcoin-mining-work. Accessed 12 December 2021c.

Blanqué, P. (2021). Money and its velocity matter: the great comeback of the quantity equation of money in an era of regime shift. Amundi Asset management. Discussion Paper 52. https://research-center.amundi.com/article/money-and-its-velocity-matter-great-comeback-quantity-equation-money-era-regime-shift. Accessed 02 March 2022.

Blanqué, P. (2019). *Essays in positive investment management* (3rd ed.). Economica.

CCAF. (2022). Cambridge Bitcoin electricity consumption index. Cambridge Centre for Alternative Finance, Cambridge. https://ccaf.io/cbeci/index. Accessed 02 February 2022.

Chaikin, A. (2012). Is SpaceX changing the rocket equation? Smithonian Magazine. https://www.smithsonianmag.com/air-space-magazine/is-spacex-changing-the-rocket-equation-132285884/. Accessed 02 February 2022.

Clarke, A. C. (1951). *The exploration of space*. Temple Press.

CoinMarketCap. (2021). All cryptocurrencies. CoinMarketCap. https://coinmarketcap.com/all/views/all/. Accessed 12 December 2021.

Curdia, V., & Woodford, M. (2011). The central-bank balance sheet as an instrument of monetary policy. *Journal of Monetary Economics*, *58*(1), 54–79. https://www.sciencedirect.com/science/article/abs/pii/S0304393210001224?via%3Dihub. Accessed 02 February 2021.

Davies, G. (2002). A history of money: From ancient times to the present day (3rd ed.).

De Vries, A., & Stoll, C. (2021). Bitcoin's growing e-waste problem. Resources Conservation and Recycling. https://doi.org/10.1016/j.resconrec.2021.105901. Accessed 02 February 2022.

Economist. (2016). Saudi Arabia adopts the Gregorian calendar. The Economist. https://www.economist.com/middle-east-and-africa/2016/12/15/saudi-arabia-adopts-the-gregorian-calendar. Accessed 02 February 2021.

Economist. (2022). The charm of cryptocurrencies for white supremacists. The Economist. https://www.economist.com/united-states/2022/02/05/the-charm-of-cryptocurrencies-for-white-supremacists. Accessed 22 February 2022.

FED. (2022). Central Bank digital currency—Frequently asked questions. Federal Reserve. https://www.federalreserve.gov/cbdc-faqs.htm. Accessed 02 February 2022.

FED. (2021). Recent balance sheet trends: Assets. Federal Reserve Board. https://www.federalreserve.gov/monetarypolicy/bst_recenttrends.htm. Accessed 02 February 2022.

FED-BOD. (2020). Federal Reserve issues FOMC statement. Federal Reserve Board of Directors. Press Release. https://www.federalreserve.gov/newsevents/pressreleases/monetary20200315a.htm. Accessed 02 February 2021.

FED-BNY. (2020). Statement regarding treasury securities, agency mortgage-backed securities, and agency commercial mortgage-backed securities operations. Federal Reserve Bank of New York. https://www.newyorkfed.org/markets/opolicy/operating_policy_200610. Accessed 02 February 2021.

Ferguson, N. (2008). *The ascent of money: A financial history of the world*. Penguin Books.

Futron Corporation. (2002). Space transportation costs: Trends in price per pound to orbit 1990–2000. Futron Corporation. https://www.yumpu.com/en/document/read/36996100/space-transportation-costs-trends-in-price-per-pound-to-orbit-/4. Accessed 02 February 2022.

Greene, B. (2004). *The fabric of the Cosmos: Space, time, and the texture of reality*. Penguin Books.

Goswami, N., & Garretson, P. (2020). *Scramble for the skies: The great power competition to control the resources of outer space*. Lexington Books.

HSBC-BCG. (2021). Delivering net zero supply chains: The multi-trillion dollar key to beat climate change. Boston Consulting Group and HSBC. https://www.hsbc.com/-/files/hsbc/news-and-insight/2021/pdf/211026-delivering-net-zero-supply-chains.pdf?download=1. Accessed 02 February 2022.

IEA. (2021). Net zero by 2050: A roadmap for the global energy sector. International Energy Agency. https://iea.blob.core.windows.net/assets/deebef5d-0c34-4539-9d0c-10b13d840027/NetZeroby2050-ARoadmapfortheGlobalEnergySector_CORR.pdf. Accessed 02 February 2022.

Jessop, S., & Shalal, A. (2021). COP26 coalition worth $130 trillion vows to put climate at heart of finance. Reuters News. https://www.reuters.com/business/cop/wrapup-politicians-exit-cop26-130tn-worth-financiers-take-stage-2021-11-03/. Accessed 02 February 2022.

John, A., Shen, S., & Wilson, T. (2021). China's top regulators ban crypto trading and mining, sending bitcoin tumbling. Reuters Business. https://www.reuters.com/world/china/china-central-bank-vows-crackdown-cryptocurrency-trading-2021-09-24/. Accessed 02 February 2022.

Lewis, J. S. (2011). "Demandite" and resources in space. John Lewis. http://www.johnslewis.com/2011/01/demandite-and-resources-in-space.html. Accessed 02 March 2022.

Lewis, J. S. (1996). *Mining the sky: Untold riches from the asteroids, comets, and planets*. Addison Wesley Publishing Company.

Malys, S., Seago, J. H., Pavlis, N. K., Seidelmann, P. K., & Kaplan, G. H. (2015). Why the Greenwich meridian moved. *Journal of Geodesy, 89*, 1263–1272. https://link.springer.com/article/10.1007/s00190-015-0844-y. Accessed 02 February 2021.

McKinsey GI. (2022). The net-zero transition What it would cost, what it could bring. Mckensey Global Institute. https://www.mckinsey.com/~/media/mckinsey/business%20functions/sustainability/our%20insights/the%20net%20zero%20transition%20what%20it%20would%20cost%20what%20it%20could%20bring/the%20net-zero%20transition-report-january-2022-final.pdf. Accessed 14 February 2022.

McLeay, M., Radia, A., & Thomas, R. (2014a). Money in the modern economy: An introduction. Bank of England. Quarterly Bulletin. https://www.bankofengland.co.uk/-/media/boe/files/quarterly-bulletin/2014a/money-in-the-modern-economy-an-introduction.pdf. Accessed 06 June 2020.

McLeay, M., Radia, A., & Thomas, R. (2014b). Money creation in the modern economy. Bank of England. Quarterly Bulletin. https://www.bankofengland.co.uk/-/media/boe/files/quarterly-bulletin/2014b/money-creation-in-the-modern-economy. Accessed 06 June 2020.

Metcalf, T., & Morales, A. (2021). Carney unveils $130 trillion in climate finance commitments. Bloomberg News. https://www.bloomberg.com/news/articles/2021-11-02/carney-s-climate-alliance-crests-130-trillion-as-pledges-soar.

Murray, B. (2021). Supply chains need $100 trillion to hit Climate Goal. Bloomberg. https://www.bloomberg.com/news/articles/2021-10-28/supply-chains-need-100-trillion-to-hit-climate-goal-paper-says. Accessed 02 February 2022.

NASA. (2021). Advanced space transportation program: Paving the highway to space. National Aeronautics Administration Agency. https://www.nasa.gov/centers/marshall/news/background/facts/astp.html. Accessed 02 February 2022.

NASA. (2018). Parker solar probe becomes fastest ever spacecraft. NASA. https://blogs.nasa.gov/parkersolarprobe/2018/10/29/parker-solar-probe-becomes-fastest-ever-spacecraft/. Accessed 02 February 2022.

Olick, D. (2021). Foreclosures are surging now that Covid mortgage bailouts are ending, but they're still at low levels. CNBC. https://www.cnbc.com/2021/10/14/foreclosures-surge-67percent-as-covid-mortgage-bailouts-expire.html. Accessed 02 February 2022.

Papazian, A. V. (2021). Sustainable finance and the space value of money. Cambridge Judge Business School. University of Cambridge. https://www.jbs.cam.ac.uk/insight/2021/sustainable-finance-and-the-space-value-of-money/. Accessed 12 December 2021.

Papazian, A. V. (2017). The space value of money. Review of financial markets (Vol. 12, pp. 11–13). Chartered Institute for Securities and Investment. Available at: http://www.cisi.org/bookmark/web9/common/library/files/sironline/RFMJan17.pdf. Accessed 16 December 2020.

Papazian, A. V. (2015). Value of money in spacetime. The International Banker. https://internationalbanker.com/finance/value-of-money-in-spacetime/. Accessed 02 February 2021.

Papazian, A. V. (2013a). Space exploration and money mechanics: An evolutionary challenge. International Space Development Hub. https://papers.ssrn.com/sol3/papers.cfm?abstract_id=2388010 Accessed 06 June 2021.

Papazian, A. V. (2013b). Our financial imagination and the Cosmos. Cambridge Judge Business School. University of Cambridge. https://www.jbs.cam.ac.uk/insight/2013b/our-financial-imagination-and-the-cosmos/. Accessed 12 December 2021.

Papazian, A. V. (2013c). Economics and finance: The frontlines of galactic evolution. *Principium, 7*, 8–9. https://i4is.org/wp-content/uploads/2017/01/Principium_7_NovDec_2013.pdf. Accessed 02 February 2022

Papazian, A. V. (2011a) A product that can save a system: Public capitalisation notes. In Collected Seminar Papers (2011a–2012) (Vol. 1). Chair for Ethics and Financial Norms. Sorbonne University.

Papazian, A. V. (2011b). Islamic money creation (Vol. 8, pp. 19–21). Islamic Banking and Finance. http://www.iefpedia.com/english/wp-content/uploads/2011b/11/IBF28ArmenPapazian.pdf. Accessed 02 February 2022.

Prasad, E. (2021). Five myths about cryptocurrency. The Washington Post. https://www.washingtonpost.com/outlook/five-myths/cryptocurrency-yths-bitcoin-dogecoin-musk/2021/05/20/1f3f6c28-b8ad-11eb-96b9-e949d5397de9_story.html. Accessed 02 February 2022.

Ranasinghe, D. (2021). Global Debt is fast approaching record $300 trillion-IIF. Reuters Business. https://www.reuters.com/business/global-debt-is-fast-approaching-record-300-trillion-iif-2021-09-14/. Accessed 02 February 2021.

Rovelli, C. (2018). *The order of time*. Riverhead Books.

Smolin, L. (2006). *The trouble with physics*. Penguin Books.

Space X. (2022). Capabilities and services. Space X. https://www.spacex.com/media/Capabilities&Services.pdf. Accessed 02 March 2022.

Treasury Direct. (2021). Monthly statement of the public debt of the United States November 30, 2021. US Department of the Treasury Bureau of the Fiscal Service. https://www.treasurydirect.gov/govt/reports/pd/mspd/2021/opdm112021.pdf. Accessed 02 February 2022.

UNFCCC. (2021). COP 26 and the Glasgow Financial Alliance for Net Zero (GFANZ). https://racetozero.unfccc.int/wp-content/uploads/2021/04/GFANZ.pdf. Accessed 02 February 2022.

UNFCC. (2021). UN climate chief urges countries to deliver on USD 100 billion pledge. UN Climate Change. https://unfccc.int/news/un-climate-chief-urges-countries-to-deliver-on-usd-100-billion-pledge. Accessed 02 February 2022.

US Treasury. (2021). Debt limit. US Treasury. https://home.treasury.gov/policy-issues/financial-markets-financial-institutions-and-fiscal-service/debt-limit. Accessed 02 February 2022.

9

Conclusion

Look again at that dot. That's here. That's home. That's us. On it everyone you love, everyone you know, everyone you ever heard of, every human being who ever was, lived out their lives. The aggregate of our joy and suffering, thousands of confident religions, ideologies, and economic doctrines, every hunter and forager, every hero and coward, every creator and destroyer of civilization, every king and peasant, every young couple in love, every mother and father, hopeful child, inventor and explorer, every teacher of morals, every corrupt politician, every 'superstar', every 'supreme leader', every saint and sinner in the history of our species lived there–on a mote of dust suspended in a sunbeam.

Image Credit NASA, Voyager 1, 1990

The Earth is a very small stage in a vast cosmic arena. Think of the rivers of blood spilled by all those generals and emperors so that, in glory and triumph, they could become the momentary masters of a fraction of a dot. Think of the endless cruelties visited by the inhabitants of one corner of this pixel on the scarcely distinguishable inhabitants of some other corner, how frequent their

A. V. Papazian, *The Space Value of Money*,
https://doi.org/10.1057/978-1-137-59489-1_9

misunderstandings, how eager they are to kill one another, how fervent their hatreds.

Our posturings, our imagined self-importance, the delusion that we have some privileged position in the Universe, are challenged by this point of pale light. Our planet is a lonely speck in the great enveloping cosmic dark. In our obscurity, in all this vastness, there is no hint that help will come from elsewhere to save us from ourselves.

The Earth is the only world known so far to harbor life. There is nowhere else, at least in the near future, to which our species could migrate. Visit, yes. Settle, not yet. Like it or not, for the moment the Earth is where we make our stand.

It has been said that astronomy is a humbling and character-building experience. There is perhaps no better demonstration of the folly of human conceits than this distant image of our tiny world. To me, it underscores our responsibility to deal more kindly with one another, and to preserve and cherish the pale blue dot, the only home we've ever known.

Carl Sagan, Pale Blue Dot, 1994[1]

As humanity continues to pursue the targets of the Paris Agreement (UNFCCC, 2015), to keep world temperature increases below 2 °C above pre-industrial levels and ideally limit the temperature increase to 1.5 °C, sustainable finance will soon become the only type of finance tolerated across the planet. Indeed, as a direct result of the Paris Agreement, the Global SDGs, the Net Zero initiative, and the raging climate crisis, sustainability and sustainable finance have become a mainstream agenda, ushering in changes in financial regulation, disclosure requirements, value chains, and business strategies. This sustainability transformation has triggered the development of new standards, new frameworks, and tools that aim to support the transition to a sustainable Net Zero economy by mid-century (CBD, 2021; CPI, 2021; Dasgupta, 2021; EU, 2022; IEA, 2021; IFRS, 2021; IPCC, 2022; IPBES, 2019; McKinsey GI, 2022; TCFD 2017, 2021a, 2021b; TNFD 2021; UNEPFI, 2021).

As encouraging as these developments are, the rise of sustainable finance is yet to penetrate the analytical content and value framework of core finance theory and practice—a framework built entirely on risk and time. As we develop the frameworks of sustainable finance, we must come to the realisation that what is really needed and currently missing is a framework of value built around responsibility. Indeed, we have littered every environment

[1] Sagan (1994) is referring to the photograph of Earth taken Feb. 14, 1990, by NASA's Voyager 1 at a distance of 3.7 billion miles (6 billion kilometres) from the Sun. https://solarsystem.nasa.gov/resources/536/voyager-1s-pale-blue-dot/.

we have come to touch. The plastic in the oceans, the carbon in the air, the waste in our rivers and on land, and the debris in orbit are the undeniable evidence that we need to, and must, reinvent human productivity. Our financial value paradigm and analytical framework are key and central elements of this monumental task. To achieve a successful transition, our financial mathematics of value and return must reflect our commitment to a sustainable future.

The principle and equations proposed in this book aim to entrench sustainability into finance, making finance inherently sustainable. The proposed framework identifies the need to redefine the value of money beyond risk and time, by integrating space into core finance theory and practice—as an analytical dimension and our physical context, stretching from within matter to inside the planet, its surface, atmosphere, and outer space. Once space is introduced into financial analysis, alongside the risk-averse investor, the holy grail of the field since inception, two key stakeholders become relevant to the logic of the value of money: a pollution-averse planet and an aspirational human society.

I propose the introduction of a missing principle of value, i.e., the Space Value of Money, that can facilitate the theoretical and mathematical adjustment of our models. A principle that entrenches our spatial responsibility into our value framework (Papazian, 2011, 2013a, b, c, 2015, 2017, 2021). The space value of money complements our existing principles and facilitates the multidimensional assessment and valuation of cash flows.

Indeed, risk and return and time value of money cater to the mortal risk-averse investor without any reference to her/his/their responsibilities vis-à-vis space, vis-à-vis the planet and humanity. Moreover, given their internal biases, these principles and associated equations have inadvertently discriminated against our evolutionary investments, i.e., investments with high risks and very distant returns. Indeed, our current time and risk focused value framework leaves human evolutionary investments in a blind spot.

The space value of money principle requires that a dollar invested in space has at the very least a dollar's worth of positive impact on space. It requires that our mathematical expressions of value and return reflect this fact. Such that, alongside risk and time, the value of cash flows and assets are also defined and assessed vis-à-vis space, by their space impact.

The principle establishes the lower threshold of acceptable investments. While investors are free to pursue their positive returns, they should not achieve them at the expense of the planet and/or humanity. The principle requires that we avoid negative impacts and optimise positive ones through the space value metrics, like Gross and Net Space Value, which quantify the planetary, human, and economic impact of investments. These key aspects of impact are further defined as Pollution and Biodiversity impacts, Human Capital and R and D impacts, and New Asset and New Money impacts. The space value metrics I propose can help quantify the space impact of an investment across all the layers of space it affects, and offer a possible logic through which we can integrate impact into our value and return equations and our decision making.

Transforming the logic of the value of money by integrating space, as an analytical dimension and our physical context, entrenches responsibility and sustainability into our models. This leads to a transformed optimisation target across markets and investments: we must optimise space impact while we minimise risks and maximise returns. In turn, this changes our financial mathematics of value and return in sync with our new sustainability objectives.

Furthermore, the principle and metrics provide a framework that can be used to design and build the sustainable finance algorithms of the future. By providing the necessary theoretical logic and mathematical equations, and thus elements and data points needed for space impact quantification and integration, the space value framework facilitates the design and deployment of relevant trading algorithms that can help and support the growth and expansion of sustainable finance within the context of already highly digitised capital markets.

The space value of money principle and framework can also help us transform our monetary architecture. If investors must have a positive space impact, so must money creators. Our current money creation methodology, i.e., debt-based money, imposes several limitations on our potential in space. Due to the logic and calendar time-linked nature of debt instruments, debt-based money chains our species to an artificially created imaginary calendar, acting like monetary gravity that chains us to the surface of the planet. Meanwhile, given that money is continuously created via debt, in any economy, there always is a large number of households, businesses, corporations, banks, and government institutions forced to chase banknotes and deposits in order to fulfil their debt obligations—a monetary hunger that stimulates growth on the one hand, but also consumes our planet unnecessarily.

I propose an alternative approach to money creation built upon the introduction of a new monetisation instrument, Public Capitalisation Notes. PCNs are designed as high positive space impact, non-debt, no maturity, equity-like, risk and asset sharing instruments that can help us fund the transition to Net Zero. I call the approach Value Easing, an approach that is far *less* inflationary than its precedents, Credit and Quantitative Easing, as it injects new money directly into productive investments where they are needed most. Indeed, PCNs and Value Easing can be used to address all our evolutionary challenges including outer space exploration.

The transition to Net Zero, which is nothing short of an evolutionary leap that involves reinventing human productivity, requires that we rethink the value of money, and the principles, equations, and institutional arrangements that govern its creation, allocation, and deployment.

On the path to a sustainable human civilisation, towards true and effective transformation, the most serious and persistent challenges will be found not in our technological imagination, where we will surely witness the emergence of incredible new technologies, but in our financial imagination, where time and risk have intricately bonded with our institutional architecture and value models, ignoring space and omitting our responsibility of impact on it and in it.

By formally introducing space into our financial value framework, as an analytical dimension and our physical context, by entrenching our responsibility for space impact into finance, by changing the logic of the value of money to account for risk, time, and space parameters, we may be able to facilitate the needed long-term transformations in human productivity, ensuring both its sustainability and continuous expansion across time and space.

References

CBD. (2021). A New Global Framework for Managing Nature Through 2030: 1st Detailed Draft Agreement Debuts. Convention on Biological Diversity. https://www.un.org/sustainabledevelopment/blog/2021/07/a-new-global-framework-for-managing-nature-through-2030-1st-detailed-draft-agreement-debuts/. Accessed 02 February 2022.

CPI. (2021). Global landscape of climate finance 2021. Climate Policy Initiative. https://www.climatepolicyinitiative.org/wp-content/uploads/2021/10/Full-report-Global-Landscape-of-Climate-Finance-2021.pdf. Accessed 02 February 2022.

Dasgupta, P. (2021). The economics of biodiversity: The Dasgupta Review. HM Treasury. https://assets.publishing.service.gov.uk/government/uploads/system/uploads/attachment_data/file/962785/The_Economics_of_Biodiversity_The_Dasgupta_Review_Full_Report.pdf. Accessed 02 March 2021.

EU. (2022). EU taxonomy for sustainable activities. European Commission. https://ec.europa.eu/info/business-economy-euro/banking-and-finance/sustainable-finance/eu-taxonomy-sustainable-activities_en. Accessed 12 February 2022.

IEA. (2021). Net zero by 2050: A roadmap for the global energy sector. International Energy Agency. https://iea.blob.core.windows.net/assets/deebef5d-0c34-4539-9d0c-10b13d840027/NetZeroby2050-ARoadmapfortheGlobalEnergySector_CORR.pdf. Accessed 02 February 2022.

IFRS. (2021). IFRS Foundation announces International Sustainability Standards Board. https://www.ifrs.org/news-and-events/news/2021/11/ifrs-foundation-announces-issb-consolidation-with-cdsb-vrf-publication-of-prototypes/. Accessed 02 February 2022.

IPBES. (2019). The global assessment report on Biodiversity and Ecosystem Services. Intergovernmental Science-Policy Platform on Biodiversity and Ecosystem Services. https://ipbes.net/system/files/2021-06/2020%20IPBES%20GLOBAL%20REPORT%28FIRST%20PART%29_V3_SINGLE.pdf Accessed 02 February 2021.

IPCC. (2022). Climate change 2022: Impacts, adaptation and vulnerability. Summary for policymakers. Intergovernmental Panel on Climate Change. https://report.ipcc.ch/ar6wg2/pdf/IPCC_AR6_WGII_SummaryForPolicymakers.pdf. Accessed 28 February 2022.

McKinsey GI. (2022). The net-zero transition What it would cost, what it could bring. Mckensey Global Institute. https://www.mckinsey.com/~/media/mckinsey/business%20functions/sustainability/our%20insights/the%20net%20zero%20transition%20what%20it%20would%20cost%20what%20it%20could%20bring/the%20net-zero%20transition-report-january-2022-final.pdf. Accessed 14 February 2022.

McLeay, M., Radia, A., & Thomas, R. (2014a). Money in the modern economy: An introduction. Bank of England. Quarterly Bulletin. https://www.bankofengland.co.uk/-/media/boe/files/quarterly-bulletin/2014a/money-in-the-modern-economy-an-introduction.pdf. Accessed 06 June 2020.

McLeay, M., Radia, A., & Thomas, R. (2014b). Money creation in the modern economy. Bank of England. Quarterly Bulletin. https://www.bankofengland.co.uk/-/media/boe/files/quarterly-bulletin/2014money-creation-in-the-modern-economy. Accessed 06 June 2020.

Papazian, A. V. (2011). A product that can save a system: Public capitalisation notes. In Collected Seminar Papers (2011–2012) (Vol. 1). Chair for Ethics and Financial Norms. Sorbonne University.

Papazian, A. V. (2013a). Space exploration and money mechanics: An evolutionary challenge. International Space Development Hub. https://papers.ssrn.com/sol3/papers.cfm?abstract_id=2388010. Accessed 06 June 2021.

Papazian, A. V. (2013b). Our financial imagination and the Cosmos. Cambridge Judge Business School. University of Cambridge. https://www.jbs.cam.ac.uk/insight/2013b/our-financial-imagination-and-the-cosmos/. Accessed 12 December 2021.

Papazian, A.V. (2013c). Economics and finance: The frontlines of galactic evolution. *Principium, 7*, 8–9. https://i4is.org/wp-content/uploads/2017/01/Principium_7_NovDec_2013.pdf. Accessed 02 April 2022.

Papazian, A. V. (2015). Value of money in Spacetime. The International Banker. https://internationalbanker.com/finance/value-of-money-in-spacetime/. Accessed 02 February 2021.

Papazian, A. V. (2017). The space value of money. Review of financial markets (Vol. 12, pp. 11–13). Chartered Institute for Securities and Investment. Available at: http://www.cisi.org/bookmark/web9/common/library/files/sironline/RFMJan17.pdf Accessed 16 December 2020.

Papazian, A. V. (2021). Sustainable finance and the space value of money. Cambridge Judge Business School. University of Cambridge. https://www.jbs.cam.ac.uk/insight/2021/sustainable-finance-and-the-space-value-of-money/. Accessed 02 February 2022.

Sagan, C. (1994). *Pale blue dot*. Random House.

Smolin, L. (2006). *The trouble with physics*. Penguin Books.

TCFD. (2021a). Forward looking financial metrics consultation. Task Force on Climate-related Financial Disclosures. https://assets.bbhub.io/company/sites/60/2021a/03/Summary-of-Forward-Looking-Financial-Metrics-Consultation.pdf. Accessed 02 February 2022.

TCFD. (2021b). Proposed guidance on climate-related metrics, targets, and transition plans. Task Force on Climate-related Financial Disclosures. https://assets.bbhub.io/company/sites/60/2021b/05/2021b-TCFD-Metrics_Targets_Guidance.pdf. Accessed 01 January 2022.

TCFD. (2017). Final report: Recommendations of the task force on climate-related financial disclosures. https://assets.bbhub.io/company/sites/60/2020/10/FINAL-2017-TCFD-Report-11052018.pdf. Accessed 02 February 2021.

TNFD. (2021). Taskforce on nature-related financial disclosures. https://tnfd.global/. Accessed 02 February 2022.

UNEPFI. (2021). UNEPFI webpage on principles of responsible banking. https://www.unepfi.org/banking/bankingprinciples/. Accessed 02 Febraury 2022.

UNFCCC. (2015). Paris agreement. United Nations Framework Convention on Climate Change. https://unfccc.int/sites/default/files/english_paris_agreement.pdf. Accessed 02 December 2020.

Index

A. V. Papazian, *The Space Value of Money*,
https://doi.org/10.1057/978-1-137-59489-1